Lecture Notes in Computer Science 6941

Commenced Publication in 1973
Founding and Former Series Editors:
Gerhard Goos, Juris Hartmanis, and Jan van Leeuwen

Pamela Forner   Julio Gonzalo
Jaana Kekäläinen   Mounia Lalmas
Maarten de Rijke (Eds.)

Multilingual and Multimodal Information Access Evaluation

Second International Conference
of the Cross-Language Evaluation Forum, CLEF 2011
Amsterdam, The Netherlands, September 19-22, 2011
Proceedings

 Springer

Volume Editors

Pamela Forner
Center for the Evaluation of Language and Communication Technologies (CELCT)
Via alla Cascata 56/c, 38123 Povo, Italy
E-mail: forner@celct.it

Julio Gonzalo
National University of Distance Education, E.T.S.I. Informática de la UNED
c/Juan del Rosal 16, 28040 Madrid, Spain
E-mail: julio@lsi.uned.es

Jaana Kekäläinen
University of Tampere, School of Information Sciences
Kanslerinrinne 1, 33014 Tampere, Finland
E-mail: jaana.kekalainen@uta.fi

Mounia Lalmas
Yahoo! Research
Avinguda Diagonal 177, 8th Floor, 08018 Barcelona, Spain
E-mail: mounia.lalmas@gmail.com

Maarten de Rijke
University of Amsterdam, Intelligent Systems Laboratory
Science Park 107, 1098 XG Amsterdam, The Netherlands
E-mail: derijke@uva.nl

ISSN 0302-9743   e-ISSN 1611-3349
ISBN 978-3-642-23707-2   ISBN 978-3-642-23708-9 (eBook)
DOI 10.1007/978-3-642-23708-9
Springer Heidelberg Dordrecht London New York

Library of Congress Control Number: 2011935602

CR Subject Classification (1998): I.2.7, H.2.8, I.7, H.3-5, H.5.2, I.1.3

LNCS Sublibrary: SL 3 – Information Systems and Application, incl. Internet/Web
and HCI

Typesetting: Camera-ready by author, data conversion by Scientific Publishing Services, Chennai, India

Printed on acid-free paper

Springer is part of Springer Science+Business Media (www.springer.com)

Preface

Since 2000 the Cross-Language Evaluation Forum (CLEF) has played a leading role in stimulating research and innovation in a wide range of key areas in the domain of information retrieval. It has become a landmark in the annual research calendar of the international information retrieval and search community. Through the years, CLEF has promoted the study and implementation of evaluation methodologies for diverse types of retrieval task and search scenario. As a result, a broad, strong and multidisciplinary research community has been created, which covers and spans the different areas of expertise needed to deal with the evaluation of solutions to challenging information tasks.

Until 2010, the outcomes of experiments carried out under the CLEF umbrella were presented and discussed at annual workshops in conjunction with the European Conference for Digital Libraries. CLEF 2010 represented a radical departure from this "classic" CLEF format. While preserving CLEF's traditional core business and goals, namely, benchmarking activities carried in various tracks, we complemented these activities with a peer-reviewed conference component aimed at advancing research in the evaluation of complex information systems for cross-language tasks and scenarios. CLEF 2010 was thus organized as an independent four-day event consisting of two main parts, a peer-reviewed conference followed by a series of laboratories and workshops.

CLEF 2011 continued to implement this new format, with keynotes, contributed papers and lab sessions, but with a few additional refinements. First, in this year's event we interleaved the conference presentations and the laboratories over a three-and-a-half-day period. Second, we added a "community session" aimed at creating awareness of funding opportunities, offering networking opportunities, and demonstrating emerging infrastructures that help support cross-language retrieval experiments.

This year, the papers accepted for the conference included research on evaluation methods and settings, natural language processing within different domains and languages, multimedia and reflections on CLEF. Two keynote speakers highlighted important developments in the field of evaluation. Elaine Toms (University of Sheffield) focused on the role of users in evaluation, whereas Omar Alonso (Microsoft Corporation) presented a framework for the use of crowdsourcing experiments in the setting of retrieval evaluation.

CLEF 2011 featured six benchmarking activities: ImageCLEF, PAN, CLEF-IP, QA4MRE, LogCLEF and, new this year, MusiCLEF. In parallel, there was a workshop dedicated to the evaluation of information access technologies in the setting of cultural heritage. All the experiments carried out by systems during the evaluation campaigns are described in a separate publication, namely, the Working Notes, distributed during CLEF 2011 and available on-line.

The community session at CLEF 2011 was organized around a strategic EU meeting, a networking session organized by the Chorus Network of Excellence, an Evaluation Initiatives session with overviews from other benchmarking fora, and an infrastructure session dedicated to the DIRECT system for handling data resulting from retrieval experiments.

The success of CLEF 2011 would not have been possible without the invaluable contributions of all the members of the Program Committee, Organizing Committee, students and volunteers that supported the conference in its various stages. Thank you all!

Finally, we would like to express our gratitude to the sponsoring organizations for their significant and timely support.

These proceedings were prepared with the assistance of the Center for the Evaluation of Language and Communication Technologies (CELCT), Trento, Italy.

July 2011

Pamela Forner
Julio Gonzalo
Jaana Kekäläinen
Mounia Lalmas
Maarten de Rijke

Organization

CLEF 2011 was organized by the Intelligent Systems Lab Amsterdam (ISLA) of the University of Amsterdam.

General Chairs

Julio Gonzalo   National University of Distance Education (UNED), Spain

Maarten de Rijke   University of Amsterdam, The Netherlands

Program Chairs

Jaana Kekäläinen   University of Tampere, Finland

Mounia Lalmas   Yahoo! Research Barcelona, Spain

Labs Chairs

Paul Clough   University of Sheffield, UK

Vivien Petras   Humboldt University of Berlin, Germany

Resource and Publicity Chair

Khalid Choukri   Evaluations and Language resources Distribution Agency (ELDA), France

Organization Chair

Emanuele Pianta   Center for the Evaluation of Language and Communication Technologies (CELCT), Italy

Organizing Committee

Intelligent Systems Lab Amsterdam (ISLA), University of Amsterdam, The Netherlands:

Richard Berendsen
Bouke Huurnink
Edgar Meij
Maarten de Rjke
Wouter Weerkamp

Center for the Evaluation of Language and Communication Technologies (CELCT), Italy:

Pamela Forner
Giovanni Moretti

Program Committee

Pia Borlund	Royal School of Library and Information Science, Denmark
Pavel Braslavski	Yandex, Russia
Ben Carterette	University of Delaware, USA
Nicola Ferro	University of Padua, Italy
Norbert Fuhr	University of Duisburg, Germany
Kal Järvelin	University of Tampere, Finland
Gareth Jones	Dublin City University, Ireland
Marko Junkkari	University of Tampere, Finland
Noriko Kando	National Institute of Informatics, Japan
Heikki Keskustalo	University of Tampere, Finland
Birger Larsen	Royal School of Library and Information Science, Denmark
Prasenjit Majumder	Dhirubhai Ambani Institute of Information and Communication Technology, India
Paul McNamee	Johns Hopkins University, USA
Stefano Mizzarro	University of Udine, Italy
Jian-Yun Nie	University of Montreal, Canada
Douglas Oard	University of Maryland, USA
Carol Peters	ISTI CNR Pisa, Italy
Vivien Petras	Humboldt University, Germany
Ian Ruthven	University of Strathclyde, UK
Tetsuya Sakai	Microsoft Research Asia
Christa Womser-Hacker	University of Hildesheim, Germany

Sponsoring Institutions

CLEF 2011 benefited from the support of the following organizations:

Center for Creation, Content and Technology, University of Amsterdam, The Netherlands
City of Amsterdam, The Netherlands
ELIAS Research Network Programme
European Language Resources Association, France
Microsoft Research
Netherlands Institute for Sound and Vision, Hilversum, The Netherlands
Netherlands Organization for Scientific Research, The Netherlands
PROMISE Network of Excellence
University of Amsterdam, The Netherlands
Xerox Research Centre Europe, France

Table of Contents

Visual and Context

"Would You Trust Your IR System to Choose Your Date?" Re-thinking IR Evaluation in the 21st Century

Elaine G. Toms

Information School, University of Sheffield
Regent Court, Portobello, Sheffield, UK
e.toms@sheffield.ac.uk

Abstract. This talk examines interactive IR system evaluation from the holistic approach, including some of the pitfalls in existing approaches, and the issues involved in designing more effective processes and procedures.

Keywords: User-centred evaluation, interactive information retrieval.

1 Overview

In the quest for almost any concept or object, human nature is surprisingly and paradoxically predictable: people will either know it when they see it, or they know what it should be before they find it. Yet, the concept of the "right" document, photo or indeed any information object is in the mind of the seeker regardless of whether the seeker is looking for dinner, a new car, a date, a novel to read, or a solution to a work problem. To complicate the matter, that solution is also a moving target, and not necessarily a single reality out there to be uncovered. Once an object is found, whether it is pertinent, relevant, accurate or correct is a *human* judgment made at a particular moment in time for a particular purpose. An information retrieval (IR) system can only provide suggestions; its role is to support and facilitate, and not to replace the human decision-making process.

In the evaluation of interactive IR systems, this requires a directional shift from the typical TREC, INEX and even CLEF evaluation processes. At present, evaluation has moved from an emphasis on topical relevance, to an emphasis on measuring almost anything that can be quantified. This is more likely to be data extracted from transaction logs in an attempt to develop a more predictable personalized search with likely the same accuracy as forecasting the future by reading tealeaves in a cup!

We have failed to step back and assess the broader picture. What exactly are we evaluating and for what purpose? It could be said that we have focused far too long on the tool and not on what the tool will be used for. For example, if that IR system was choosing ones date, a restaurant or a medical treatment, would we use the same evaluation techniques in use today? This talk will consider evaluation from that holistic and contextual perspective. It will examine some of the pitfalls in existing approaches, and discuss the issues involved in designing more effective evaluation approaches for assessing interactive IR systems.

P. Forner et al. (Eds.): CLEF 2011, LNCS 6941, p. 1, 2011.

Crowdsourcing for Information Retrieval Experimentation and Evaluation

Omar Alonso

Microsoft Corp.
Mountain View, California, USA
omar.alonso@microsoft.com

Abstract. Very recently, crowdsourcing has emerged as a viable alternative for conducting different types of experiments in a wide range of areas. Generally speaking and in the context of IR, crowdsourcing involves outsourcing tasks to a large group of people instead of assigning such tasks to an employee or editor. The availability of commercial crowdsourcing platforms offers vast access to an on-demand workforce. This new approach makes possible to conduct experiments extremely fast, with good results at a low cost. However, like in any experiment, there are several implementation details that would make an experiment work or fail. For large scale evaluation, deployment in practice is not that simple. Tasks have to be designed carefully with special emphasis on the user interface, instructions, content, and quality control.

In this invited talk, I will explore some directions that may influence the outcome of a task and I will present a framework for conducting crowdsourcing experiments making some emphasis on a number of aspects that should be of importance for all sorts of IR-like tasks. Finally, I will outline research trends around human computation that promise to make this emerging field even more interesting in the near future.

P. Forner et al. (Eds.): CLEF 2011, LNCS 6941, p. 2, 2011.
© Springer-Verlag Berlin Heidelberg 2011

Building a Cross-Language Entity Linking Collection in Twenty-One Languages

James Mayfield[1], Dawn Lawrie[1,2], Paul McNamee[1], and Douglas W. Oard[1,3]

[1] Johns Hopkins University Human Language Technology Center of Excellence
[2] Loyola University Maryland
[3] University of Maryland, College Park

Abstract. We describe an efficient way to create a test collection for evaluating the accuracy of cross-language entity linking. Queries are created by semi-automatically identifying person names on the English side of a parallel corpus, using judgments obtained through crowdsourcing to identify the entity corresponding to the name, and projecting the English name onto the non-English document using word alignments. We applied the technique to produce the first publicly available multilingual cross-language entity linking collection. The collection includes approximately 55,000 queries, comprising between 875 and 4,329 queries for each of twenty-one non-English languages.

Keywords: Entity Linking, Cross-Language Entity Linking, Multilingual Corpora, Crowdsourcing.

1 Introduction

Given a mention of an entity in a document and a set of known entities, the *entity linking* task is to find the entity ID of the mentioned entity within a knowledge base (KB), or return NIL if the mentioned entity was previously unknown. In the *cross-language entity linking* task, the document in which the entity is mentioned is in one language (e.g., Serbian) while the set of known entities is described using another language (in our experiments, English). Entity linking is a crucial requirement for automated knowledge base population, and can be used to generate metadata about entities that can be used to improve search tasks.

Entity linking has been the subject of significant study over the past five years. Pioneering work focused on matching entity mentions to Wikipedia articles [5,7]. Although focused on clustering equivalent names rather than entity linking, the ACE 2008 workshop conducted evaluations of cross-document entity coreference resolution in Arabic and English [4] but not across languages. In 2009, the Text Analysis Conference (TAC) Knowledge Base Population track (TAC KBP) conducted a formal evaluation of English entity linking using a fixed set of documents and Wikipedia articles [11]. Shared tasks with a variety of characteristics have since emerged elsewhere, including CLEF [2], FIRE [15], and NTCIR.[1] Very recently, TAC[2] and NTCIR have both for the first time defined a shared task for cross-language entity linking.

---

[1] http://ntcir.nii.ac.jp/CrossLink/
[2] http://nlp.cs.qc.cuny.edu/kbp/2011/

P. Forner et al. (Eds.): CLEF 2011, LNCS 6941, pp. 3–13, 2011.
© Springer-Verlag Berlin Heidelberg 2011

Over the years, CLEF has included two tasks that called for some aspects of entity linking capabilities. The first was the WebCLEF task in 2005 and 2006 [3], which focused on known-item search, such as finding a named Web page. For example, given the query "El Palacio de la Moncloa" (Moncloa Palace), a system should return the URL: http://www.lamoncloa.gob.es/. More recently, the Web People Search (WePS) task in CLEF 2010 [2] extended this task to also include the extraction of attributes for different people that could be referred to by the same name. In this paper, by contrast, we focus on linking references to people for whom we already have attributes available (by using Wikipedia infoboxes as our knowledge base). Moreover, we construct those references in the context of a document rather than as isolated queries because we are ultimately interested in extracting attributes from documents once they have been linked.

The goals of this work are to identify a way to efficiently create cross-language entity linking training and test data, and to apply that method to create such collections in many languages. We hope by doing this to accelerate the identification of the best methods for performing cross-language entity linking; to foster entity linking research by researchers who have interest in specific languages beyond the few languages that TAC and NTCIR will soon support; and to promote the development of language-neutral approaches to cross-language entity linking that will be applicable to many of the world's languages.

Our approach to collection creation has two distinguishing characteristics: the use of parallel document collections to allow most of the work to occur in a single language, and the use of crowdsourcing to quickly and economically generate many human judgments. A fundamental insight on which our work is based is that if we build an entity linking test collection using the English half of a parallel text collection, we can make use of readily available annotators and tools developed specifically for English, then project the English results onto the other language. Thus, we apply English named entity recognition (NER) to find person names in text, an English entity linking system to identify candidate entity IDs, and English annotators to select the correct entity ID for each name. We use standard statistical word alignment techniques to map from name mentions in English documents to the corresponding names in non-English documents. Projection of named entity annotations is known to be imperfect [17]; we therefore intend to use crowd-sourcing again to curate the name projections.

The increasing availability of multi-way parallel text collections offers the potential for further leverage, allowing the same ground truth English annotations to be projected to more than one language. We demonstrate this capability on three multi-way parallel text collections that together cover 18 non-English languages, plus (to extend the diversity of character sets) single-pair parallel text collections for Arabic, Chinese, and Urdu. Moreover, by building several test collections from the same English annotations, we expect our test collection and other collections built in this way to enable work on comparative analysis of cross-language entity linking effectiveness across languages.

2 Document Collections

To support the goal of creating a cross-language person-entity linking test collection, a parallel collection should include a large amount of text that is rich in person names, at least some of which refer to well-known people (because publicly-available knowledge bases tend to be populated with well-known entities). Large collections are not typically required for entity linking experiments. However, we want to support not just evaluation but also training of machine learning-based linkers, as well as the automated learning of translation and/or transliteration models; there is little value in providing evaluation resources for new languages if we cannot provide for training and development testing as well. We therefore need a sufficient number of names for training, development and evaluation partitions. The parallel text collections shown in Table 1 meet these requirements.[3] Together, these collections contain 196,717 non-English documents in five different scripts.

Table 1. Our sources of parallel text

Collection	Obtained from
Arabic	LDC (LDC2004T18)
Chinese	LDC (LDC2005T10)
Europarl5	http://www.statmt.org/europarl/
ProjSynd	http://www.statmt.org/wmt10/
SETimes	http://elx.dlsi.ua.es/~fran/SETIMES/
Urdu	LDC (LDC2006E110)

An alternative to using parallel texts would have been to link from Wikipedia, which of course contains multilingual content and cross-language links for equivalent names. Wikipedia text is, however, highly stylized and thus not as representative of the broad range of naturally occurring documents that we would like to link from as we would wish.

3   Choosing Names

To identify person names in the English side of each parallel text collection we used the publicly available named entity recognition system created by Ratinov and Roth [13], for which the highest published score on the CoNLL 2003 dataset has been reported [16]. That resulted in 257,884 unique person name/document pairs across the six collections. We then eliminated all single-token names. Because named entity recognition is imperfect, we manually examined these English results to eliminate strings that were obviously not person names. We also eliminated names that occurred only once across the collection, and we limited to ten the number of times a single name string would be included (to avoid building a collection dominated by a small number of common names). We used person names exclusively in this collection; however, building test collections for other entity types, such as organizations, could be handled in exactly the same way.

---

[3] We also considered using the EMEA and EMILLE corpora, but felt that they had inadequate coverage of person names. The recently-released Europarl v6 should permit the inclusion of several additional Central and Eastern European languages.

4   Generating English Ground Truth

The HLTCOE submitted competitive results in the TAC 2009 and 2010 entity linking evaluations, and Recall@3 for our system on the non-NIL PER subset of the TAC 2010 collection (i.e., those queries that match a Wikipedia entry) was over 94% (201/213). Only one additional correct answer would have been added by increasing the depth beyond 3; all other correct answers were missing entirely from the entity linker output. Accordingly we felt comfortable using the HLTCOE entity linking system [12] to create a ranked list of candidate entities from the TAC KBP knowledge base,[4] and presenting only the top three entries to human judges.

We expect the Recall@3 to be higher in this collection for two reasons. First, approximately half of our recall errors on the TAC KBP data were single-word names; for this paper we use only full names as English queries. Second, the TAC 2010 queries were selected to ensure a substantial number of confusable names [9]; for this paper we made no effort to restrict the collection to names that are highly ambiguous. Therefore, we believe the percentage of queries in the collection that are incorrectly categorized as having no matching Wikipedia page to be under 1%.

We collected human judgments using Amazon's Mechanical Turk [1], which has been applied to a wide array of HLT problems [6,14]. A paid assessor, called a 'Turker,' could select one of the three candidates, "None of the above" (if none of the three was the correct referent), "Not a person" (indicating an NER error) or "Not enough information." A single Mechanical Turk Human Intelligence Task (HIT) consisted of six such sets, two of which were interleaved queries for which we already knew ground truth. The three candidates were shown in a random order, rather than the best-first order returned by the entity linker.

Measures were collected to ensure quality assessments by the Turkers. The average time required to complete a HIT (i.e., six queries) was 2.5 minutes, with only 14 of 314 Turkers averaging under one minute per HIT. No assessments were eliminated due to insufficient time spent on the task. In addition, we examined assessments of ground truth. More than half of the Turkers answered all ground truth questions correctly (647 of 1139[5]). An accuracy score was computed based on the fraction of correct assessments a Turker made on ground truth queries within each batch. This score became more finely tuned with each task a Turker submitted. Assessors with an accuracy below 80% were eliminated from the judgment pool. The mean accuracy over all the Turkers was 94.6% with a standard deviation of 11.3%. A total of 81 Turkers were eliminated for accuracy below 80%; however, the number of HITs submitted by these Turkers was very low. Most poor performers only sampled the task, then apparently moved on to other opportunities.

We obtained three separate judgments for each query, and included the query in the collection only if none of the three Turkers had been eliminated for low accuracy and only if all three Turkers agreed on the answer. The loss of queries is reported in

---

[4] The TAC knowledge base is derived from an October 2008 subset of Wikipedia pages that contained Infoboxes; more than 114k persons are represented in the KB.

[5] Many Turkers provided annotations for multiple batches. Performance on ground truth was calculated for each batch submitted to Mechanical Turk.

Table 2. Fraction of all person names lost as queries during the human assessment phase

Reason for Attrition	Queries Lost
Low Turker quality	0.9%
Turker disagreement	0.9%
Missing judgments	0.3%

Table 2. On average we paid about US$0.28 to obtain multiple annotations for each English query, which works out to about US$0.08 per non-English query across the 21 languages.

5 Name Projection

Name projection involves several steps, which are taken to improve quality and ensure that an exact string match of the name is found in the document. First, the Berkeley Word Aligner [8] creates a mapping from words in the English text to words or phrases in the target language. Second, for each name identified by the NER system, a contiguous span of tokens in the target language document is associated with the name. This is based on the assumption that all names are written sequentially in each target language. This approach can compensate for the aligner missing the middle of a name, especially when the middle portion appears only in the non-English document. By aligning all names, rather than only those in the query set, the entire collection can be used to compensate for a bad alignment in a particular document. The third step ranks all the projections for a single English name based on frequency. In the final step, the most frequent target language string appearing in the target language document is chosen as the projection for the English query name.

As an example of this process, consider the query "Tony Blair." Suppose the English document is searched for occurrences of "Tony Blair," which is found to align with the single Arabic word for "Blair" (بلير). By using this projection alone, the target language query would become بلير ("Blair"). However, by using all alignments across the entire collection, we find that the most frequent alignment of "Tony Blair" is the Arabic translation of his full name توني بلير ("Tony Blair"). If the query document also contains the Arabic "Tony Blair," which aligns to the English "Blair," the full name will be chosen for the query.

To determine the quality of automated name projection, we took a convenience sample of five of the twenty-one languages. An assessor familiar with the sampled language[6] performed manual assessment of each sampled projection. The outcome of this assessment is shown in Figure 1. In four of the languages we evaluated a random sample of 100 name projections; for Spanish we manually evaluated the entire query set. The results show that the proportion of fully correct projections varies from just under 70% to 98%. In addition we observe that less than 10% of the queries are completely

---

[6] A native speaker, or in some cases a non-native speaker with years of college study.

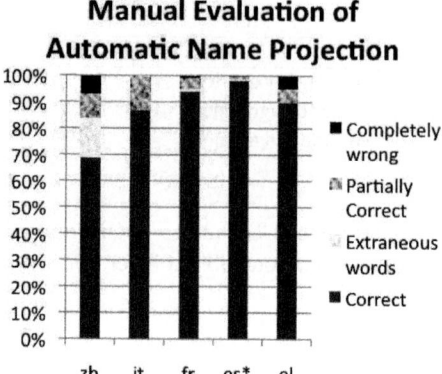

Manual Evaluation of Automatic Name Projection

Fig. 1. Accuracy of automated name projection as determined by manual inspection. * indicates that all the queries were inspected. In other cases a random sample of 100 queries was inspected.

wrong for any of the languages evaluated. In particular, the proportion of completely wrong queries drops to 1% or less for roman script languages. The evaluators also indicated whether the entire name was embedded in the non-English query string (labeled "Extraneous words" in Figure 1) or whether at least one word associated with the name appeared in the non-English string (labeled "Partially Correct").

To obtain more accurate name projections, human assistance is required. Such assistance is well suited to crowd-sourcing as it is relatively simple for a bilingual speaker to view the set of paired sentences in the parallel documents that contain the English name and find the name within the target language sentences. However, such a process is not free, and all 55,000 queries require name projection. The above evaluation demonstrates that a large majority of the projections are correct. To reduce curation costs, we would like to automatically identify a subset of queries that with high probability have the correct name projections. We used Google Translate[7] (which supports all of our languages) to automatically translate each projected name back into English. If the resulting name translation exactly matches the original English query names, the projected name might reasonably be considered correct, and no further curation need be performed. The results for this process by language appear in Figure 2. On average, about 72% of the projected names translate back to exact matches. There is a high variability in this result across the languages; Chinese exhibits the lowest exact match rate at 14%, while Romanian has the highest rate at 91%. In addition to identifying exact matches, we also searched the translated string for the query name to determine if the only problem with the translation was extraneous words. This condition was only found in about 2% of the queries. In about 20% of the queries, partial matches were present where at least one of the translated words was found in the query. 7% of the queries exhibited no words in common. Figure 2 also shows the proportion of query names that failed to project.

---

[7] http://translate.google.com/

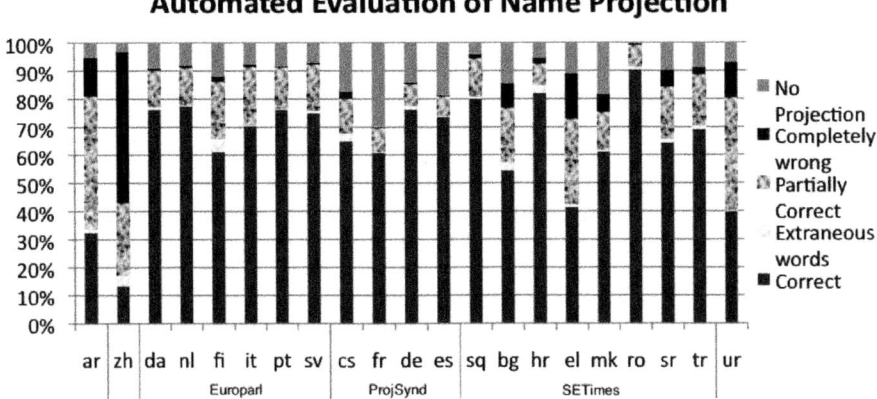

Fig. 2. Automated evaluation of name projection when Google Translate is used to translate the projected queries back into English

A significant portion of these are the result of missing documents or missing sentences in the target language documents. However, there is no doubt that some of the failed projections could be recovered manually. Finally, the discrepancy between the correctness found through manual inspection and automated translation is in many cases due to minor differences in spelling or the inclusion of an accented character.

Automatically identifying name projections that are correct with high likelihood presents great opportunities for cost savings. By accepting these projections as correct, only 28% of queries require human assessment.

6 Collection Statistics

One desirable characteristic of an entity linking test collection is balance between the number of NIL queries (i.e., those for which no resolution can be made) and non-NIL queries; detecting that an entity cannot be resolved is an important requirement in many entity linking applications.[8] Table 3 shows that this goal was well met.

The NER system originally identified 257,884 English person names across the six parallel collections. Not all of these names end up as queries; significant attrition occurs in an effort to maintain collection quality. The various sources of query attrition, together with the percentage of the person names lost for each, are shown in Table 4. Some of these forms of attrition could be ameliorated to increase the collection size. A total of 14,806 English queries resulted from our procedure. These correspond to 59,224 queries across the 21 languages. Further attrition caused by projecting the English names onto those twenty-one languages as shown in Table 2 resulted in a final non-English query count of 55,157.

---

[8] Serbian can be written in both Latin and Cyrillic alphabets; our collection uses the Latin alphabet.

Table 3. Language coverage in our collection

Language	Collection	Queries	Non-NIL
Arabic (ar)	Arabic	2,829	661
Chinese (zh)	Chinese	1,958	956
Danish (da)	Europarl	2,105	1,096
Dutch (nl)	Europarl	2,131	1,087
Finnish (fi)	Europarl	2,038	1,049
Italian (it)	Europarl	2,135	1,087
Portuguese (pt)	Europarl	2,119	1,096
Swedish (sv)	Europarl	2,153	1,107
Czech (cs)	ProjSynd	1,044	722
French (fr)	ProjSynd	885	657
German (de)	ProjSynd	1,086	769
Spanish (es)	ProjSynd	1,028	743
Albanian (sq)	SETimes	4,190	2,274
Bulgarian (bg)	SETimes	3,737	2,068
Croatian (hr)	SETimes	4,139	2,257
Greek (el)	SETimes	3,890	2,129
Macedonian (mk)	SETimes	3,573	1,956
Romanian (ro)	SETimes	4,355	2,368
Serbian (sr)[9]	SETimes	3,943	2,156
Turkish (tr)	SETimes	3,991	2,169
Urdu (ur)	Urdu	1,828	1,093
Total		55,157	29,500

Table 4. Fraction of all person names lost as queries due to various factors during the query creation phase

Reason for Attrition	Queries Lost
Single-word name	45.1%
More descriptive name appears in document	1.1%
Manual name curation	5.0%
Only one occurrence of name in collection	15.8%
Ten occurrences of name already included	11.6%
Could not locate name in English document	0.5%
To avoid predicted NIL/non-NIL imbalance	4.0%

7   Using the Test Collection

To determine whether the collection is suitable for training cross-language entity linking systems, we built cross-language entity linkers for several of the languages. Our approach to *monolingual* entity linking breaks the problem into two phases: (1) identification of a relatively small set of plausible KB entities, or *candidate identification*; and (2) ranking of those candidates using supervised machine learning (*candidate ranking*). The ranking step orders the candidates, including NIL, by the likelihood that each is a correct match for the query entity.

To identify candidate entities we rely on a number of quickly calculable name comparisons. We create indexes to support rapid identification of (1) KB entries with an exact name match; (2) entities with an alternative name that matches the query (e.g., *Duchess of Cambridge* for *Catherine Middleton*); (3) entities with name fragments (given names or surnames) in common with the query; and (4) entities sharing character 4-grams with the query entity. This candidate identification phase provides a three to four orders of magnitude reduction in the number of entities to which our full battery of comparison features must be applied.

We rank candidates using a ranking support vector machine (SVM^{rank}) [10]. Feature vectors representing candidate alignments to KB entries include features based on name similarity, textual context, matches of relations found in the KB, named entities that occur in both the KB and the query document, and indications of absence from the knowledge base. A more complete description of our entity linking system can be found in McNamee et al.[12].

To construct cross-language entity linkers, we augmented our monolingual entity linking system with features based on transliteration (for name matching) and cross-language information retrieval (for matching terms surrounding the name mention against terms found on the candidate's Wikipedia page). Table 5 shows that cross-language entity linking accuracy is nearly as good as English-only entity linking using the same (unprojected) queries for Arabic, German and Spanish, but for Bulgarian and Greek cross-language entity linking is considerably harder. The Bulgarian and Greek results comport with our intuition that low-resource language pairs that require transliteration pose additional challenges for cross-language entity linking. Moreover, the good cross-language results for several language pairs suggest (at least for those languages) that name projection errors resulting from incorrect alignments are not a large source of measurement error.

Table 5. Success@1 on development partition, English-only vs. Cross-language

Language	English	Cross-Language	% Monolingual
Arabic	0.9192	0.9131	99.3%
Bulgarian	0.9873	0.8456	85.6%
German	0.9283	0.9148	98.5%
Greek	0.9776	0.7852	80.3%
Spanish	0.9317	0.9366	100.5%
Turkish	0.9857	0.9402	95.4%

8   Conclusion

We have demonstrated a methodology for creating cross-language entity linking test collections, and used that methodology to create collections in twenty-one languages. We also described challenges in creating this first multilingual cross-language entity linking test collection, such as the quality of crowdsourced judgments and bilingual projection of query names. We showcased the utility of the collection with preliminary cross-language entity linking results. Our approach uses existing aligned parallel corpora; this decision allows exploitation of existing high-quality English tools to economically obtain cross-language entity linking annotations. The test collection is available at http://web.jhu.edu/HLTCOE/datasets.html.

References

1. Amazon.com: Amazon Mechanical Turk (2005),
   https://www.mturk.com/mturk/welcome
2. Artiles, J., Borthwick, A., Gonzalo, J., Sekine, S., Amigo, E.: Overview of the web people search clustering and attribute extraction tasks. In: CLEF Third WEPS Evaluation Workshop (2010)
3. Balog, K., Azzopardi, L., Kamps, J., Rijke, M.D.: Overview of WebCLEF 2006. In: Cross-Language Evaluation Forum, pp. 803–819 (2006)
4. Baron, A., Freedman, M.: Who is Who and What is What: Experiments in cross-document co-reference. In: Proceedings of the 2008 Conference on Empirical Methods in Natural Language Processing, pp. 274–283. Association for Computational Linguistics, Honolulu (2008), http://www.aclweb.org/anthology/D08-1029
5. Bunescu, R.C., Pasca, M.: Using encyclopedic knowledge for named entity disambiguation. In: European Chapter of the Assocation for Computational Linguistics, EACL (2006)
6. Callison-Burch, C., Dredze, M.: Creating speech and language data with Amazon's Mechanical Turk. In: Proceedings of the NAACL HLT 2010 Workshop on Creating Speech and Language Data with Amazon's Mechanical Turk, CSLDAMT 2010, pp. 1–12. Association for Computational Linguistics, Stroudsburg (2010), http://portal.acm.org/citation.cfm?id=1866696.1866697
7. Cucerzan, S.: Large-scale named entity disambiguation based on Wikipedia data. In: Empirical Methods in Natural Language Processing (EMNLP) (2007)
8. Haghighi, A., Blitzer, J., DeNero, J., Klein, D.: Better word alignments with supervised itg models. In: Proceedings of the Joint Conference of the 47th Annual Meeting of the ACL and the 4th International Joint Conference on Natural Language Processing of the AFNLP, ACL 2009, vol. 2, pp. 923–931. Association for Computational Linguistics, Stroudsburg (2009), http://portal.acm.org/citation.cfm?id=1690219.1690276
9. Ji, H., Grishman, R., Dang, H.T., Griffitt, K., Ellis, J.: Overview of the TAC 2010 Knowledge Base Population track. In: Text Analysis Conference, TAC (2010)
10. Joachims, T.: Training Linear SVMs in Linear Time. In: Proceedings of the 12th ACM SIGKDD International Conference on Knowledge Discovery and Data Mining, KDD 2006, pp. 217–226. ACM, New York (2006),
    http://doi.acm.org/10.1145/1150402.1150429
11. McNamee, P., Dang, H.T.: Overview of the TAC 2009 Knowledge Base Population track. In: Proceedings of the Text Analysis Conference (2009)
12. McNamee, P., Dredze, M., Gerber, A., Garera, N., Finin, T., Mayfield, J., Piatko, C., Rao, D., Yarowsky, D., Dreyer, M.: HLTCOE Approaches to Knowledge Base Population at TAC 2009. In: Proceedings of the Text Analysis Conference (2009)
13. Ratinov, L., Roth, D.: Design challenges and misconceptions in named entity recognition. In: Proceedings of the Thirteenth Conference on Computational Natural Language Learning (CoNLL 2009), pp. 147–155. Association for Computational Linguistics, Boulder (2009), http://www.aclweb.org/anthology/W09-1119

14. Snow, R., O'Connor, B., Jurafsky, D., Ng, A.Y.: Cheap and fast—but is it good?: evaluating non-expert annotations for natural language tasks. In: Proceedings of the Conference on Empirical Methods in Natural Language Processing, EMNLP 2008, pp. 254–263. Association for Computational Linguistics, Stroudsburg (2008), http://portal.acm.org/citation.cfm?id=1613715.1613751
15. Tiwari, C.: et al.: News Assist: Identifying set of relevant entities used in news article. In: FIRE (2010)
16. Tjong Kim Sang, E., Meulder, F.D.: Introduction to the CoNLL-2003 shared task: Language-independent named entity recognition. In: Conference on Natural Language Learning, CoNLL (2003)
17. Yarowsky, D., Ngai, G.: Inducing multilingual POS taggers and NP bracketers via robust projection across aligned corpora. In: NAACL (2001)

Search Snippet Evaluation at Yandex: Lessons Learned and Future Directions

Denis Savenkov, Pavel Braslavski, and Mikhail Lebedev

Yandex
16, Leo Tolstoy St., Moscow 119021
{denxx,pb,mlebedev}@yandex-team.ru

Abstract. This papers surveys different approaches to evaluation of web search summaries and describes experiments conducted at Yandex. We hypothesize that the complex task of snippet evaluation is best solved with a range of different methods. Automation of evaluation based on available manual assessments and clickthrough analysis is a promising direction.

Keywords: evaluation, snippets, search summaries, web search, experimentation.

1 Introduction

A list of ranked document summaries is de facto a standard for web search result representation. A search summary[1] commonly consists of document title, original document fragments (namely text *snippets*), and metadata such as document date, size, URL, etc. Now we can observe the tendency of enriching web search summaries with images, so called QuickLinks, links to maps (e.g. in case the retrieved document is a company or organization homepage), user ratings of different kind, and other clues. Most text snippets originate from the original document and contain highlighted terms from the initial user query or their derivatives. Some snippets are, in fact, manually-crafted summaries from third-party sites (such as ODP[2] descriptions) or from META field of the original HTML page. A wide use of *microformats*[3] shifts the emphasis from the methods of choosing the best fragments from the original text to deciding whether to use the semantic mark-up provided by the page owner or not.

In some cases summaries can provide the user with the required information in situ (e.g. factoid questions). However, the main purpose of a search summary is to inform the user about the degree of relevance of the original retrieved document. Many studies confirm that search summaries have a big impact on the perceived search quality of search: the user is unlikely to click on a misleading summary of a relevant

---

[1] Also referred to as *result summary, snippet, query-biased summary, caption,* and *document surrogate.*
[2] http://dmoz.org/
[3] http://microformats.org/

P. Forner et al. (Eds.): CLEF 2011, LNCS 6941, pp. 14–25, 2011.

document and, conversely, the user will be disappointed by a non-relevant document, if the summary suggested the opposite (however, the latter is a much less critical case). Turpin et al. [18] investigated how accounting for summary judgment stage can alter IR systems evaluation and comparison results. Based on a small user study, authors estimate that "14% of highly relevant and 31% of relevant documents are never examined because their summary is judged irrelevant" [18].

Web summary evaluation differs from search quality evaluation for several reasons. First, the notion of a "good summary" is multifaceted and sometimes contradictory. It is often hard to balance out different requirements. E.g. a snippet containing many query terms from different fragments of the original document is, in general, less readable. Longer snippets bear more information about the retrieved document but hinder overall comprehension of the search engine results page (SERP), etc. Second, summary judgments are only partially reusable (changes in generation algorithm lead to changes in an arbitrary subset of snippets for given query-document pairs).

In the industrial settings snippet evaluation can be aimed at different goals: 1) comparison with competitors, 2) evaluation of a new versions of snippet generation algorithm against production version, and 3) evaluation in favor of machine-learned algorithms for snippet generation.

In the next section we survey different approaches to search summaries evaluation and work in related areas. Section 3 describes different techniques used for snippet evaluation at Yandex, a Russian web search engine serving about 120M queries daily: an exploratory eye-tracking experiment, manual assessment of search snippets in terms of informativity and readability, automatic metrics, and evaluation based on clickthrough mining. Section 4 concludes and outlines the directions for further research.

2   Related Work

Snippet generation can be seen as a variant of general summarization task. There are two main approaches to summarization evaluation: 1) comparison against a gold standard or 2) task-oriented evaluation. Additionally, some intrinsic aspects of summaries such as readability or grammaticality are evaluated. Concurrent comparison, or side-by-side evaluation, of several summary variants is another option.

There are some approaches implemented within a series of standalone experiments or within evaluation campaigns of a larger scale.

In their pioneering work Tombros & Sanderson [17] compared the utility of query-biased summaries against first few sentences of retrieved documents in search results. A user study with 20 participants was performed on TREC *ad hoc* track data, i.e. topics and judged documents. Precision and recall of relevance judgments on summaries vs. leading sentences compared to available full document judgments, speed of judgments, and the need to refer to the full text were the indicators of the search results representation quality.

The task-oriented approach by White et al. [20] is in principal similar to one by Tombros & Sanderson. However, they tried to make search tasks closer to a real-world scenario and obtain a richer feedback from the users. 24 participants in the user study were asked to complete different search tasks using four different web search systems. Researchers used detailed questionnaires, accompanied by think-aloud, informal discussions, and automatic logging of users' actions during the experiment. The questionnaires contained the following statements regarding summary quality to be rated by participants: *The abstracts/summaries helped me to assess the pages for relevance; The abstracts/summaries showed my query terms in context.* The main automatic measure was the time spent on tasks.

Eye-tracking is a promising technique for testing user interfaces, including search results representation. Eye-tracking was used for investigation how snippet length affected user performance on navigational and informational search tasks [4]. The main findings are that longer snippets improved performance for informational queries but worsened it for navigational queries. Eye-tracking allowed to support these conclusions, i.e. a longer snippet distracted the user's attention from the URL line. The study [10] supports findings that different query types are best answered by snippets of different length. Leal Bando et al. [12] used eye-tracking in a small user study (four query-document pairs, 10 participants) to juxtapose document's fragments used by humans for generative vs. extractive query-biased summaries und showed that humans focused on the same pieces of text for both tasks most of the time. Comparison of automatically generated against human-crafted snippets suggested that gold-standard evaluation must account not only for word overlap but also for position information.

Mechanical Turk[4] crowdsourcing was used in a study on temporal snippets [2]. Mechanical Turk judges, presented with three variants of snippets for a Wikipedia page at once, had to choose the best one and provide additional response. 30 snippets corresponding to 10 queries were evaluated in total.

Clarke et al. studied snippet features that potentially influenced snippet quality and consequently – user behavior [3]. The authors performed clickthrough mining of a commercial search engine. In contrast to previous work based on rather small user studies, this study enabled a large-scale experiment in a less artificial setting. The authors looked at *clickthrough inversions* as a signal of snippet attractiveness: the pairs of consequent snippets in result list, where the lower result received more clicks than the higher-ranked one. The study confirmed the perception that the presence of query terms in a snippet, its length, complexity of URL, and readability contribute to overall quality of snippets.

Kanungo & Orr [11] reported on a machine-learned readability measure for search snippets. The model was trained on about 5,000 human judgments and incorporated 13 various features such as *average characters per word, percentage of complex words, number of fragments, query word hit fraction* etc. The trained model predicted human judgment well and can be used both for continuous large-scale monitoring of snippet readability and for improving existing summarizers.

DUC/TAC series of workshops[5] has been focusing on evaluation methodology for automatic summarization for several years. The initiative collected a sizeable volume of system-produced summaries, ideal human-crafted summaries, and comparisons of system summaries with ideal summaries performed by humans. These data enabled

---

[4] http://www.mturk.com/
[5] http://duc.nist.gov/, http://www.nist.gov/tac/

the introduction of automatic quality measures based on proximity of an automatically generated summary and a set of ideal summaries. Proximity can be defined in terms of common n-grams, word sequences, or similar syntactic units. ROUGE [13] and Basic Elements (BE) [6] exemplify these approaches and show a good correlation with systems rankings based on human judgments. Automatic measure allows re-using of judgments.

The last edition of the TAC multidocument summarization included 46 topics for guided summarization. The task was to produce a 100-word summary from the first 10 documents on a certain topic and an update summary for the second 10 documents. Automatically generated summaries were evaluated and compared to ideal summaries by human judges in respect of responsiveness (relevance to topic), readability, and Pyramid (content similarity to human summaries) [14]. In contrast to web queries, the task presents a much more detailed description of the information need, its aspects, and prior knowledge on the topic.

Snippet generation can be seen as passage retrieval task, i.e. retrieving the fragments of a document relevant to a particular information need. Passage retrieval task was evaluated within TREC HARD track in 2003[5] and 2004. System results were evaluated against fragments of documents marked as relevant by annotators. How to quantify the character-level overlap of ideal fragments with systems' output is discussed in [19].

Two years (2007 & 2008) WebCLEF[6] offered snippet generation/information synthesis task: participants were presented with a topic description and up to 100 Google results to relevant search queries. A system response was a ranked list of plain text snippets extracted from the retrieved documents (first 7,000 characters of the system response were assessed). System responses were pooled, and assessors were asked to mark text spans with useful information. Average character precision and average character recall were used for evaluation similarly to TREC HARD track. ROUGE-1 and ROUGE-1-2 turned out to be not quite appropriate for evaluation of the task [9, 15].

Recently INEX announced a snippet evaluation track [7]. The task is to return snippets limited to 300 characters for retrieved Wikipedia articles. Evaluation metrics will employ comparison of relevance assessments based on whole documents vs. short snippets.

1CLICK subtask of the NTCIR-9 Intent task [1] is running at the time of writing (June 2011). It resembles snippet generation, QA, and information synthesis tasks: for a given query the system must return a string of 140 ('mobile run') or 500 ('desktop run') characters. A Japanese collection and Japanese queries are used. Evaluation is based on information nuggets presented in the system's response (similar to content similarity in TAC evaluation).

3 Snippet Evaluation at Yandex

In order to establish a snippet evaluation routine at Yandex, we experimented with a wide range of techniques and approaches in line with those described in Section 2:

---

[6] http://ilps.science.uva.nl/WebCLEF/

pairwise comparison of two versions/systems, relevance on whole documents vs. snippets, direct readability assessment, clickthrough mining, etc. The work is still in progress. Our current perception is that it is very hard to invent an integral measure of snippet quality. Thus, we suggest using a set of different tools and approaches for different aspects and goals of snippet evaluation.

3.1 Eye Tracking Experiment

Eye-tracking became very popular for investigating user behavior and usability of user interfaces. We employed eye-tracking for better understanding of how different aspects of snippet quality influence user satisfaction. One of the research questions was whether highlighting additional terms reflecting possible user intents was helpful.

We prepared 19 tasks of different types, e.g. download a given popular song, find information for writing an essay on a given topic, find the address of a given movie theatre, find term definition, etc. Some tasks were open, while for others initial search queries were provided. 20 participants took part in the study, each participant was allotted an hour to complete the tasks. Both experienced and beginner, frequent and occasional Yandex users took part in the study. Participants were divided into two groups – the first group was presented with standard snippets, the second group had snippets with terms related to the query intent (e.g. "buy" for commercial queries) highlighted along with the query terms.

The main conclusions from our user study are as follows:

1. The title is much more important than the body of the snippet. Users skip relevant results with no highlighted terms in the title in favor of lower-ranked results with seemingly better titles.
2. Highlighting attracts users' attention and helps them navigate through the results list. Users click directly on highlighted terms in the snippet titles. Additional highlighted terms, e.g. reflecting query intents, help users find the answer faster and draw their attention to results in the lower part of SERP (supports [8], see Fig.1).
3. Experienced users prefer skimming: they examine snippet fragments around highlighted words, jumping from one part of the snippet to another. If the title contains relevant information, these users prefer clicking on the link without examining the body of the snippet.
4. Users rely on ranking – high-ranked results are clicked regardless of the snippet's content or quality (supported by many click-log experiments). However, some users get bored after examining a few results at the top of the result list and scroll down to the lower part.
5. Inexperienced users are somewhat "scared to click"; they usually examine a considerable number of results before clicking. Novices examine snippet content more thoroughly before moving on to the next result.
6. Users go to the original document, even if the snippet contains a complete answer to their factoid query (supports [2]).
7. Some users are conservative and shy away from a certain type of snippets, e.g. containing image or video thumbnails.

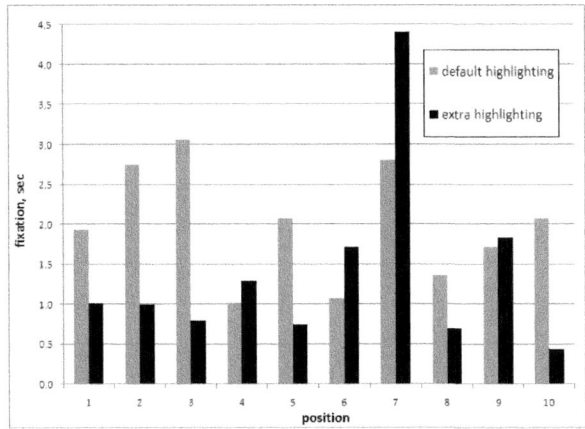

Fig. 1. Averaged fixation times for two groups (10 participants each) solving the same task: 1) default highlighting – query terms only; 2) query intents ("download", in this case) additionally set off in bold in snippets at positions 7 and 9

3.2 Manual Assessment

Manual assessment performed by trained judges is the basis of traditional evaluation methodology in the field of information retrieval. We performed a sizeable manual assessment within the experiments on machine-learned snippets.

The key features of an ideal snippet are: 1) it conveys sufficient information about the whole document in the context of a query (i.e. users can assess the document's relevance to the query based on a snippet); 2) it is easy to read/understand. These qualities are reflected in the snippet *informativity* and *readability* measures.

During initial experiments we realized that it was hard for an assessor to score a snippets' informativity on an absolute scale. Even if we have absolute scores it is questionable whether these scores are comparable across different queries. A much easier task is to compare and rank different snippet variants for a given query-document pair.

The interface of the assessment tool is presented in Fig. 2. Assessor is presented with a query and a randomly ordered list of up to 10 snippet variants for the same document produced by different snippet generation algorithms. Query-document pairs were sampled so that their relevance distribution was close to Yandex's results. The task was to move individual snippets up and down and insert "borderlines", thus, creating ranked groups of snippets of an approximately equal quality. Evaluation for informativity and readability was performed independently (different assessors ranked the same task for informativity and for readability). The study was performed in three stages; 11 judges participated in the study. Table 1 describes some statistics of the evaluation process. It is interesting to observe a learning effect in informativity evaluation in terms of speed and quality: average time spent on task decreases, as well as the proportion of tasks with at least two candidates reverse-ordered compared to tasks evaluated by assessors' supervisor. Time spent on readability evaluation does not show this behavior and rather correlates with the average snippet length.

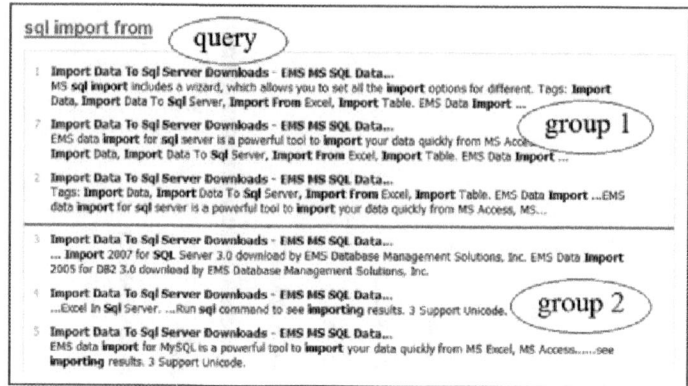

Fig. 2. Tool for relative assessment of different snippet variants for a query-document pair

Table 1. Statistics of manual evaluation experiment

Period	Query-doc pairs	Ave. snippet length	Time spent on inform. task, sec	Reverse-ordered pairs, inform., %	Time spent on readab. task, sec
Mar 2010	1,200	250	115	29	72
Jun 2010	3,200	170	107	26	60
Jan 2011	2,000	250	101	24	84

Table 2. Correlation between readability and informativity

Query length	# of query-doc pairs	# of snippets	r
1	164	1,481	0.432
2	256	2,266	0.401
3	273	2,466	0.374
4	183	1,588	0.363
≥ 5	237	2,024	0.353
total	1,113	9,825	0.383

Based on a subset of evaluation results, we calculated Kendall tau-b correlation between informativity and readability rankings (Table 2). One can see that the correlation is positive, i.e. snippets tend to be good or bad in both aspects. However, the correlation is low, which implies that we have to consider readability and informativity as complementary snippet features. It is interesting to notice that informativity and readability are less correlated in long queries. For short queries a readable text fragment containing one or two query terms is more likely to contain useful information; for longer queries this dependency becomes more complex.

It is worth to mention that the same evaluation guidelines (but different interface) were used by assessors performing "blind" side-by-side comparison of snippets on Yandex against competitors.

3.3 Automated Quality Measures

Manual assessment is very expensive and time-consuming even considering the availability of services like Mechanical Turk. When changing the snippet generation algorithm, we need a simple and fast method to assess the new version. At the moment, we use a range of automated measures that capture some snippet features:

- General number of highlighted terms, proportion of query terms presented in the snippet, proportion of highlighted terms and their variations, such as density and diversity of highlighted terms, number of highlighted terms in title, etc.
- Snippet's 'neatness', which is closely related to its readability. We measure the number of non-readable characters (#, %, ^, @, *, <, >, etc.), the number of porn words, etc.
- The number of 'empty' snippets (i.e. title-only snippets).

Table 3 presents Kendall tau-b correlation between some automated measures and manual rankings of snippets regarding informativity and readability (calculated on the same data as in Table 2).

Table 3. Correlation between assessors' rankings and rankings based on automated measures

Query length	Informativity vs.		Readability vs.	
	proportion of query terms	# of highlighted terms	proportion of non-readable chars	# of fragments
1	0.205	0.206	−0.322	−0.699
2	0.281	0.329	−0.304	−0.695
3	0.302	0.403	−0.309	−0.671
4	0.328	0.484	−0.327	−0.641
≥5	0.334	0.535	−0.323	−0.576
Total	0.274	0.424	−0.306	−0.657

Table 4. Automated measures for two snippet generation algorithms (2,000 queries, 17,009 snippets generated by each algorithm)

Measure	Alg1	Alg2
Proportion of query terms in snippets	0.762	0.774
Proportion of snippets containing all query terms	0.550	0.568
Snippet length in chars	165.76	161.59
# of highlighted query terms per snippet	3.317	3.368
Proportion of non-readable chars	0.020	0.022
Average word length	5.901	5.870

As expected, the proportion of query terms presented in a snippet and the number of highlighted terms positively correlated with informativity, whereas the proportion of non-readable characters and the number of fragments from the original document in a snippet negatively correlated with readability. However, the correlation is not strong, except for the number of fragments.

In addition, Table 4 presents some automated measures for two snippet generation algorithms produced during routine development at Yandex. In general, *Alg2* shows a better behavior, the only drawback is a slightly increased number of non-readable characters.

3.4 A/B Testing

Automatic evaluation of information retrieval systems based on user behavior is an area of active research. Automatic methods promise to make evaluation faster, cheaper, and more representative. However, despite that a plethora of data is available, the crucial problem remains interpreting these data in terms of quality.

We perform automatic evaluation of a new candidate snippet generation algorithm against the production version using A/B testing. A subset of user population is presented with search results with the same ranking but featuring different snippets. In general, we used a subset of metrics described in [16] (session-based metrics, such as *queries per session* or *reformulation rate* are not quite appropriate for snippet evaluation). However, in contrast to ranking evaluation, some metrics receive a different interpretation. For example, an increased CTR of lower positions in case of shorter snippets can indicate a positive change: the user develops a better general comprehension of SERP, whereas in case of ranking evaluation it might mean that good results are lower.

The main purpose of snippets is to help users find relevant documents on the search engine results page and avoid those that are irrelevant. Thus, the first important behavior characteristic is dwell time, i.e. the time the user spends on the web page after clicking the link on the search results page. The higher the proportion of SERP clicks with long dwell times is, the fewer documents with non-representative summaries there are in search results. Also, the less the abandonment rate (i.e. queries with no clicks on results) is, the better annotations the documents on SERP have. In addition, an increase of CTRs for the lower-ranked documents usually suggests that the snippets for top-ranked documents get less attention because they are not informative enough (cf. click inversions [3]). But this depends highly on the length of snippets, since the shorter the snippets are, the higher CTRs the lower documents have. In addition to dwell time, we need to take into account the time required to find the answer to the user's query. For example, the time it takes to make the first click is a very useful measure, which correlates with the time it takes to find the answer.

Table 5. A/B testing results for two snippet generation algorithms (*statistically significant at the 0.01 confidence level)

Measure	Alg1	Alg2
Abandoned queries, %	38.270	38.220 (−0.13%)*
Click inversions, %	6.8017	6.8212 (+0.29%)*
Long dwell times rate, %	72.5897	72.6088 (+0.026%)
Time to first click, sec	11.5274	11.5245 (−0.02%)
1st position CTR	0.3786	0.3790 (+0.10%)*
2nd position CTR	0.1631	0.1630 (−0.03%)
9th position CTR	0.0355	0.0357 (+0.42%)*
10th position CTR	0.0358	0.0360 (+0.27%)*

Table 6. A/B testing results for snippets with extra highlighting of possible user intents (*statistically significant at the 0.01 confidence level)

Measure	Default highlighting	Extra highlighting
Abandoned queries, %	40.0031	39.9052 (–0.25%)*
Click inversions, %	6.4506	6.4818 (+0.48%)*
Long dwell times rate, %	73.8379	73.7960 (–0.06%)
Time to first click, sec	11.6832	11.6638 (–0.17%)*
1st position CTR	0.3132	0.3138 (+0.19%)*
2th position CTR	0.1639	0.1645 (+0.33%)*
9th position CTR	0.0343	0.0347 (+1.11%)*
10th position CTR	0.0422	0.0424 (+0.45%)*

Table 5 presents user behavior metrics for two different snippet generation algorithms (the same as in the previous section). *Alg2* snippets were shown to 12.5% of users during two weeks (May 10–24, 2011). Since snippets generated by *Alg2* contained more query terms and were slightly shorter, we could observe increased CTRs, especially for lower positions. Due to this fact, click inversion rate increased (more attention to lower positions resulted in more click inversions). More highlighting resulted in a lower number of abandoned queries. Proportion of long (>30 sec) dwell times for *Alg2* was approximately the same as for *Alg1*. This might mean that *Alg2* generated more attractive snippets for both relevant and non-relevant documents. Since the total number of clicks on the links to relevant documents increased, we could conclude that *Alg2* generated better snippets than *Alg1*.

Table 6 shows the results of another experiment for snippet generation algorithms that differ only in the way they highlighted terms. The experiment was performed on 50% of users for two weeks. Clickthrough mining supported the results of the eye-tracking experiment; it showed that increased attractiveness of snippets resulted in higher CTRs and shorter times to first click.

4   Conclusions and Future Research

Based on our experiments we can conclude that the complex and diverse task of snippet evaluation is best solved with a range of different methods – user studies, automated measures, manual evaluation, and clicktrough mining.

Thus, we use eye-tracking when introducing changes in general SERP layout or snippet representation: snippet length, snippets enriched by video and image thumbnails, QuickLinks, and links to maps, customized snippets for recipes, hotels, forums, and products, extra highlighting, URL representation, etc.

Manual evaluation is employed for machine-learned snippet generation and comparison with competitors. We use relative quality assessments for two aspects of snippets – informativity and readability. The main drawback of the approach is that judgments cannot be re-used. However, approaches that allowed for re-using of data – e.g. ideal snippets extracted by humans – are much more costly and time-consuming and presumably show less inter-annotator agreement.

Automatic measures are suitable for fast, albeit rough, evaluation of snippet generation algorithms. We use them as regression tests for newly developed algorithms. Moreover, we plan to implement automated measures based on manual readability evaluation results (in a way similar to [11]).

A/B testing is the final step in shipping snippet generation algorithm to production.

We now plan to address the problem of building an integral snippet evaluation metrics and automation of snippet metrics based on available manual assessment results and click data analysis. In addition, we plan to conduct a manual assessment of information nuggets presented in snippets for factoid queries (analogously to DUC/TAC/1CLICK approach).

References

1. 1CLICK@NTCIR-9,
   http://research.microsoft.com/en-us/people/tesakai/
   1click.aspx
2. Alonso, O., Baeza-Yates, R., Gertz, M.: Effectiveness of Temporal Snippets. In: WSSP Workshop at the World Wide Web Conference, WWW (2009)
3. Clarke, C., Agichtein, E., Dumais, S., White, R.W.: The Influence of Caption Features on Clickthrough Patterns in Web Search. In: SIGIR 2007 (2007)
4. Cutrell, E., Guan, Z.: What Are You Looking For? An Eye-tracking Study of Information Usage in Web Search. In: CHI 2007 (2007)
5. HARD, High Accuracy Retrieval from Documents. TREC 2003 track guidelines, http://ciir.cs.umass.edu/research/hard/guidelines2003.html
6. Hovy, E.: Lin, C.-Y., Zhou, L.: Evaluating DUC 2005 Using Basic Elements. In: Fifth Document Understanding Conference (DUC), Vancouver, Canada (2005)
7. INEX 2011 Snippet Retrieval Track,
   https://inex.mmci.uni-saarland.de/tracks/snippet/
8. Iofciu, T., Craswell, N., Shokouhi, M.: Evaluating the Impact of Snippet Highlighting in Search. In: Understanding the User Workshop, SIGIR 2009 (2009)
9. Jijkoun, V., de Rijke, M.: Overview of webCLEF 2008. In: Peters, C., Deselaers, T., Ferro, N., Gonzalo, J., Jones, G.J.F., Kurimo, M., Mandl, T., Peñas, A., Petras, V. (eds.) CLEF 2008. LNCS, vol. 5706, pp. 787–793. Springer, Heidelberg (2009)
10. Kaisser, M., Hearst, M. A., Lowe, J. B.: Improving Search Results Quality by Customizing Summary Lengths. In: ACL 2008 HLT (2008)
11. Kanungo, T., Orr, D.: Predicting the Readability of Short Web Summaries. In: WSDM 2009 (2009)
12. Leal Bando, L., Scholer, F., Turpin, A.: Constructing Query-biased Summaries: a Comparison of Human and System Generated Snippets. In: IIiX 2010 (2010)
13. Lin, C.-Y.: ROUGE: A Package for Automatic Evaluation of Summaries. In: ACL 2004 Workshop: Text Summarization Branches Out, Barcelona, Spain (2004)
14. Nenkova, A., Passonneau, R. J., McKeown, K.: The Pyramid Method: Incorporating Human Content Selection Variation in Summarization Evaluation. TSLP 4(2) (2007)
15. Overwijk, A., Nguyen, D., Hauff, C., Trieschnigg, D., Hiemstra, D., de Jong, F.: On the Evaluation of Snippet Selection for WebCLEF. In: Peters, C., Deselaers, T., Ferro, N., Gonzalo, J., Jones, G.J.F., Kurimo, M., Mandl, T., Peñas, A., Petras, V. (eds.) CLEF 2008. LNCS, vol. 5706, pp. 794–797. Springer, Heidelberg (2009)

16. Radlinski, F., Kurup, M., Joachims, T.: How Does Clickthrough Data Reflect Retrieval Quality? In: CIKM 2008 (2008)
17. Tombros, A., Sanderson, M.: Advantages of Query Biased Summaries in Information Retrieval. In: SIGIR 1998 (1998)
18. Turpin, A., Scholer, F., Jarvelin, K., Wu, M., Culpepper, J.S.: Including Summaries in System Evaluations. In: SIGIR 2009 (2009)
19. Wade, C., Allan, J.: Passage Retrieval and Evaluation. Technical report, CIIR, University of Massachusetts, Amherst (2005)
20. White, R.W., Jose, J.M., Ruthven, I.: A Task-Oriented Study on the Influencing Effects of Query-Biased Summarisation in Web Searching. Information Processing and Management 39 (2003)

Towards a Living Lab for Information Retrieval Research and Development

A Proposal for a Living Lab for Product Search Tasks

Leif Azzopardi[1] and Krisztian Balog[2]

[1] School of Computing Science, University of Glasgow
Leif.Azzopardi@glasgow.ac.uk
[2] Dept. of Computer and Information Science, NTNU, Trondheim
Krisztian.Balog@idi.ntnu.no

Abstract. The notion of having a "living lab" to undertaken evaluations has been proposed by a number of proponents within the field of Information Retrieval (IR). However, what such a living lab might look like and how it might be setup has not been discussed in detail. Living labs have a number of appealing points such as realistic evaluation contexts where tasks are directly linked to user experience and the closer integration of research/academia and development/industry facilitating more efficient knowledge transfer. However, operationalizing a living lab opens up a number of concerns regarding security, privacy, etc. as well as challenges regarding the design, development and maintenance of the infrastructure required to support such evaluations. Here, we aim to further the discussion on living labs for IR evaluation and propose one possible architecture to create such an evaluation environment. To focus discussion, we put forward a proposal for a living lab on product search tasks within the context of an online shop.

1 Introduction

Evaluation is a key challenge within the field of Information Retrieval (IR) [14]. From early on in the history of IR, objective and precise ways to measure, compare and evaluate systems, methods and models have been central to the research conducted [13, 14]. The main advances have been through the dedicated efforts to form consortium that build and develop test collections, methodologies, and measures (such as CLEF, TREC and INEX). While test collection based research has been of great benefit to the IR community, allowing researchers to study a variety of task and domains, they do have a number of limitations [14]. The abstractions often lack realism, there is often no user/user model, nor any interaction [3, 9]. As such, ever more complicated measures that try to incorporate the user into the way that IR systems are evaluated have been developed [12]. However, to properly test IR systems, evaluation needs to be performed in context (i.e., with real users performing tasks using real-world applications). So one alternative that has been recently proposed is the introduction of "living labs"

P. Forner et al. (Eds.): CLEF 2011, LNCS 6941, pp. 26–37, 2011.

that involve and integrate users within the research process [7, 9]. This would, not only, enable the capture of real interaction and usage data, but also provide a context for testing and evaluating IR models, methods and systems. In Kelly et al. [9], they outline what such a lab might offer, be, and enable:

> *A living laboratory on the Web that brings researchers and searchers together is needed to facilitate ISSS [Information-Seeking Support System] evaluation. Such a lab might contain resources and tools for evaluation as well as infrastructure for collaborative studies. It might also function as a point of contact with those interested in participating in ISSS studies.*

According to Pirolli [11], having such a living lab available for research purposes would be,

> *a great attractor for scientific minds in diverse areas ranging from behavioral economics, incentive mechanisms, network theory, cognitive science, and human computer interaction...*

From discussions at the SIGIR 2009 Future Information Retrieval Evaluation workshop [7] , there was a clear desire from participants to be able to understand user information-seeking behavior in situ and the idea of a living lab as a way to do this was generally endorsed. It was also seen as a way to bridge the **data divide** within the research community, because currently interaction data is often only available to those working within organizations that provide real-world IR applications. A living lab would provide a common data repository and evaluation environment giving researchers (in particular from academia) the data required to undertake meaningful and applicable research. More generally though, a living lab has been presented not just as a platform for collaborative research, but also as a platform where users co-create the product, application or service (i.e., users are not just subjects of observation, but also part of the creation). Essentially, the users explore emerging ideas and scenarios in situ, the evaluation process is then fed back into the design of the product to further enhance their user experience[1]. While living labs have lots of appeal offering a number of opportunities and benefits, the development and implementation throws up some difficult challenges and problems which need to be overcome before such an evaluation platform can be realized.

The contributions we make in this paper are twofold. First, we propose one possible system architecture for a living lab based on a number of distinct web based services that provide a level of independence between the different parties involved in the research and development cycle (i.e., academics, commercial organizations, evaluation forums and users). While this is a rather idealized architecture of an IR focused living lab, it provides a starting point for serious discussion about how to implement such an idea. Second, we propose a living lab evaluation platform for an online shopping scenario. This scenario provides:

---

[1] The concept of living labs is attributed to Jarmo Suominen. See `http://staffnet.` `kingston.ac.uk/~ku07009/LivingLabs/PapersAndSlides/Day1RichardEnnals.` `pdf` for an explanation and some of the history regarding the concept of living labs.

(i) a novel set of search tasks, which have not received much attention in current evaluation forums, (ii) a problem where the size and scale is significantly more tractable than other tasks, such as web search i.e., an online retailer houses information on only a few thousand products, for which there is lots of rich interaction data, whereas the web contains billions of documents and large volumes of interaction data, (iii) product search data is not as problematic when it comes to privacy of the user (i.e., product search is can be made anonymous much more easily), (iv) the tasks in this scenario have direct economic implications, and (v) it provides an incentive for smaller online retailers to participate as they can benefit from research and development activity they could not otherwise afford. We hope that this work stimulates interest in the development of a living lab and leads to the creation of such an evaluation platform.

The remainder of this paper is structured as follows. In Section 2 we consider some of the potential steps or stages from standard test collections to living labs. In Section 3 we present an idealized system architecture for the development of a living lab that would facilitate a closer integration between researchers and industry. Then, in Section 4 we describe our proposal for an online shopping living laboratory. Finally, we conclude with a discussion on the benefits and challenges involved, before outlining the next steps in developing a living lab for IR research and development in Section 5.

2   From Standard Test Collections to Living Labs

There has been a number of different developments and proposals for Information Retrieval evaluation platforms. We shall briefly present the main approaches. They range from the Standard Test Collection approach (which typically adopts the Cranfield paradigm) to Fully Intergrated Living Labs. At each step the platforms become more and more application/user focused.

- **Standard Test Collection.** A testbed containing documents, topics and relevance judgments which allows for rigorous and replicable testing of methods, models and theory. Most TREC/CLEF/INEX collections are representatives of this type of test set.
- **Extended Test Collections.** A test collection augmented and extended by conducting a series of experiments that involve users. The usage and interaction data is recorded and distributed as part of the collection. The TREC Interactive track [5] and later the HARD track [1] both attempted to bring in the user into the loop. Although these tracks struggled to establish comparability between experimental sites, they were successful at highlighting the importance of users in IR research [15].
- **Simulation of Interaction.** Following on from the extended test collection, users and interaction are seeded, simulated and validated against the usage data in the extended test collection [2]. Alternatively, an abstracted task model could be developed (ranging from a simple search task to a more complex exercise that might not be solved in a single session) and researchers submit "simulated users" to perform that task [2].

- **Observational Test Centers.** Here users of an application would be logged and monitored (and depending on the setup this may be without the user's consent or knowledge). An observational test center would be able to build up a rich set of usage and interaction data (such as query logs) which could be used for research purposes.
- **Sandboxes.** A fully working application which can be modified by researchers to facilitate different configurations and permutations. IR toolkits (such as Lucene, Lemur or Terrier) may be viewed as lab or system based sandboxes, where one can experiment with varying some components. In the setting of this paper, our primary focus is on application based and human focused sandboxes; these enable researchers to vary and change various components of interest in an application. These changes can then be evaluated with users who volunteer to trial a different version of the live application.
- **Fully Integrated Living Lab.** The ideal scenario where users are not only observed, and researchers change configurations to perform experiments, but they are also part of the research process, and co-create the application or service through their usage behavior. Arguably, web search engines are already living labs, though their experimentation is performed strictly behind closed doors.

The above steps represent the continuum from system focused to application/user focused research and map to the spectrum provided by Kelly [8] where test collections are largely system focused, while living labs are on the opposite end spectrum and largely application focused. Our focus throughout this paper is on developing a fully integrated living lab, where we will primarily concentrate on the high-level design of the machinery required to facilitate a living lab[2]. In the next section we shall outline one possible system architecture to support a living lab evaluation platform, before describing how it would be applied in the context of an online shop in Section 4.

3 A System Architecture for Living Labs

In Figure 1, we outline a high level system architecture that includes test centers, sandboxes and a fully integrated living lab. The architecture is somewhat idealized consisting of four independent web based services that would cooperate together for mutual benefit. Service **A** is the web based evaluation forum that coordinates evaluation efforts among researchers and acts a broker between the live applications provided by services **B** and **D** and the research services developed, **C**. Service **B** facilitates access to the commercial web application and would provide the interaction and usage data (this vetted data is then supplied to Service **A**). Services of type **C** are the web based services that researchers develop. They interact with **A** to obtain data for the particular evaluation search task. Service **D** encompasses non-commercial applications for testing and evaluation with users out with the commercial application. User would interacts with

---

[2] For an excellent survey and practical guide on running controlled experiments within a living lab, we refer the reader to work performed by Kohavi et al [10].

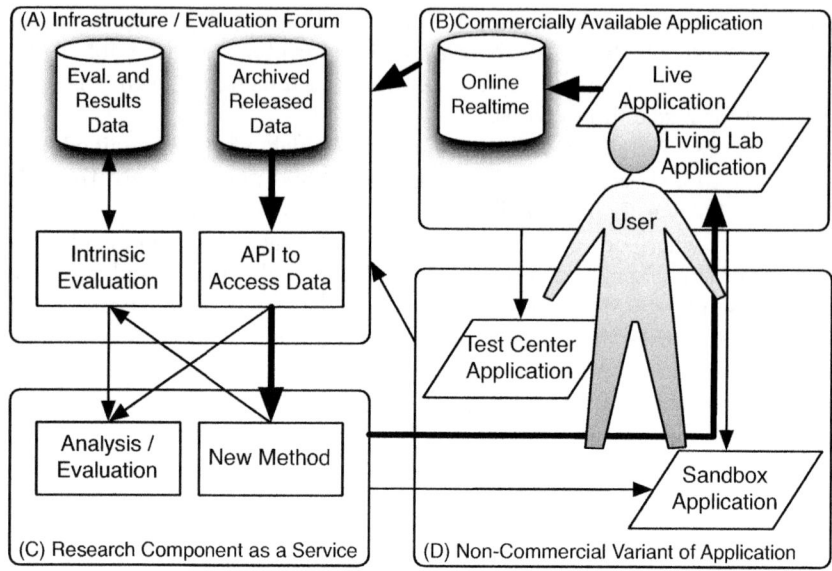

Fig. 1. A Possible System Architecture for a Living Laboratory. The thick black arrows denote the cycle of interactions required for a living lab.

the lives applications (which are denoted by diamond shapes in Figure 1); these are the Live (or Living Lab) Applications, the Test Center Applications and the Sandbox Applications. The usage data produced would be stored by service **A**, potentially along with explicit relevance data, to enable evaluations to be conducted. Next, we shall describe each service in more detail.

A) **Infrastructure/ Evaluation Forum.** This web service provides a proxy between a Live Application (**B**) and the developers and researchers of various components (**C**). It connects to **B**'s API and receives updates on the data generated by the Live App (secure one way transfer). This could happen periodically, perhaps monthly, or even continually. This Archived and Released Data repository would provide the means to perform various evaluation tasks (and might include documents, query logs, click-through data, etc.). Service **A** would also collect and collate evaluation data of different research components (which are registered with the infrastructure). It would provide two main APIs to its users, i.e., the researchers and developers of new components (**C**). One API would provide access to the data that is housed in the archive ("Data API"). The other is to provide an API to Intrinsic Evaluations that can be performed using the data within **A**. For example, this may compare the differences and similarities of results produced by the new Research Component against other existing Research Components (via a "Task API"). Being an evaluation forum/platform, Service **A** would allow Research Component Services to be registered and evaluated. If a Test

Center or Sandbox Application is used, then usage data and judgements from particular tasks could also be included within **A** to facilitate evaluation (without a living lab, or full cooperation from the providers of a live application).

B) **Commercially Available Application.** A company that delivers online services, like a search engine, online shop, etc. runs a live application, and to participate needs to supply data to **A** (once the data is vetted and moderated). End users interact with their Live Application to produce usage data. **B** may also incorporate into their application a living lab, where the live application is augmented by utilizing Research Components developed and accessible via web services of type **C** (assuming that these services can reliably and robustly handle the request and demands placed upon it by **B**). Feedback and usage data collected from the users would be collected and again exported to the evaluation forum (for the developers of **C** to analyze).

C) **Research Components as Web Services.** The developers of new components would interact with Service **A** to obtain the latest data available. A new component would use this data to perform the particular task (e.g., estimate the ranking of documents, summarize sentences, etc.). The Research Component web services would provide an API that exposes their method in a standard way for a particular task. For example, it accepts a query, and responds with a set of results in a pre-defined format. The new method developed could then be utilized by the other web services (**A**, **B** and/or **D**) as part of the Intrinsic Evaluation, Sandbox testing, or even used within a Living Lab Application.

D) **Non-Commercial Variants of Live Applications.** Here two types of applications could be created. One is a Test Center Application, which is essentially the Live Application provided by **B**, but which has been instrumented to obtain usage data (i.e., a client is created that exposes the functionality of **B** and decorates it with logging functionality). The usage data collected is exported to **A** for research and evaluation purposes. The other alternative is the Sandbox Application, where the API of **B** enables researchers to configure a variant of the Live Application to include the Research Components available through services of type **C**. Again, usage data from the Sandbox Application would be logged and provided to **A**.

These services could reside within one organization (to support in-house research), or may be distributed between the evaluation forum efforts, commercial organizations and research institutes. By breaking up the cycle into four major parts different organizations can be responsible for providing different services to facilitate research and development. This has the advantage that independence is maintained between parties. For example, the researcher of a new method can experiment and develop their algorithm without disclosing details of the algorithm, which they may wish to patent at some point. Alternatively, existing methods can be tested by invoking the API's of services of type **C** for the given task (assuming the web service is up and running) or the evaluation forum can collect evaluation results so that the performances of existing methods is available for comparison. Since researchers have access to the data, they can process the data on their own machines, using their own representations, and with their preferred programming language(s)/toolkit(s)/etc.

Commercial organizations running a live application are also buffered from potential security risks because the access to the data is via a third party/proxy. While, test centers and sandboxes can be created where users can be recruited to test variants of commercially available applications. It should be noted that test centers could be run without any direct commercial involvement. For example, companies like Bing and Google provide APIs to their search engine, so it would be quite feasible to create a web interface that connects to their search API and logs all the interactions.

While, there are advantages to separating out these concerns, this architecture does introduce a number of overheads, such as creating the services, conforming to the defined APIs for tasks, and the increased complexity in development. However, web applications are becoming substantially easier to develop, and skeleton code could be made available to help researchers expose their components as services. So far we have talked very generally about a living lab; to help focus the discussion and make the problem more tractable we shall describe how the architecture would enable evaluation of product related search tasks.

4   A Living Lab for Online Shopping

Online shopping is an activity that is commonly and frequently performed. It is attractive for customers because of the high level of convenience, broader selection, competitive pricing and greater access to information [6]. Part of the process of shopping is finding vendors, browsing and searching for products, researching products, finding reviews about products, comparing prices and buying. These tasks are performed through an intermediary search service (such as a web search engine, or portal like eBay or Amazon) and/or through direct search services provided by the online vendor. Here, we shall consider the search and browsing performed within an online shop: where the vendor's main goal is to support customers to find the products that they are interested in, the related products and similar products, to improve their online shopping experience (and ultimately to drive and increase sales).

Tasks in an Online Shopping Environment. Let us imagine an online toy shop which has a large catalogue of products. We would like to be able to: (1) let customers find the products easily through a site-search component by (a) providing some query assistance, and (b) a good ranking of products that are "relevant" to the user given the query; (2) when a customer is viewing a product provide product recommendations such as displaying related and/or similar products. For example, Patrick visits this online toy shop and would like to purchase a remote control helicopter. He queries for "RC helicopters," the system provides a number of suggestions "RC helicopters valkyre," "RC helicopters apache," "RC helicopters parts," etc., where he chooses the first suggestion. The system then returns a set of products relevant and related to this query (i.e., a number of valkyre helicopter versions and models, perhaps the competing helicopter of the same type, and commonly purchased add-ons such as batteries and blades). Patrick then selects a recent model. The system displays the web page for this product, which contains information about the product, price, ratings, etc. as well as related products such as batteries, body kits, blades, etc. for the currently selected product. Along on the page, similar products to the valkyre helicopter might also be displayed such as the apache helicopter, and other competing versions.

From this scenario there are three main search and recommendation tasks which are common to most online shops:

Query suggestion. We differentiate between two types of query suggestion. *Auto completion* refers to the functionality that recommends queries (displayed often in a drop-down list) as the user types in the search phrase; the feature is usually activated automatically after a certain number of characters are entered. *Query recommendation* is presented along with search results and offers alternative formulations of the original query; typical examples include spell correction and related searches. Displaying query recommendations is optional; as shown in sponsored search advertising, it is acceptable, and occasionally even desirable, not to show any suggestions [4].

Product search. The ability to search for products is a basic functionality that is essential for the ease and convenience of online shopping. Following the practice of web search engines, it is common to provide users with a single text input field (basic search). Many sites offer the option of advanced search, where users may put in additional filtering or matching criterion. Given the information need entered in either basic or advanced form, the search result page returned in response presents a ranked list of products, typically, along with the total number of hits and controls for paging between multiple pages of records.

Product recommendation. We distinguish between two product recommendation exercises. *Recommending similar products* is the task of offering various alternatives for a given product, which typically are displayed on the product's page. *Recommending related products* is the task of finding products that might be purchased along with the goods already selected by the user; such recommendations might be presented on any page, including product and category pages, search results (separated from organic results), and even the homepage. Product recommendations can be based on search keywords, similar items, cross-sell (related products), and up-sell (higher priced products).

How would the scenario of product search fit with the proposed architecture? Below we describe how each service might look or act given the system architecture described in Section 3.

A) Infrastructure / Evaluation Forum. The API for data would enable researchers to obtain: (1) product information, (2) usage and query log data, (3) anonymized user information, and potentially (4) trading logs. The high-level functionality for each data source is as follows.

Table 1. Search and recommendation tasks in an online shopping environment

Tasks	Inputs	Outputs
Basic product search	keyword query	ranked list of products
Query auto-completion	keyword query	ranked list of query suggestions
Query recommendation	keyword query	ranked list of query suggestions (or none)
Similar products	current product	ranked list of products
Related products	current product	ranked list of products (or none)
Product Recommendation	previous products	ranked list of products

- *Product information*: access to the list and properties of products and product categories. Certain product attributes are common to all webshops (such as name, description and price), while others can be specific to the commercial segment or to the given vendor. The product description, ratings and perhaps even product reviews might also be included.
- *Usage and query log*: number of times a query was issued, follow-up queries, search results clicked, number of times a product/category page was visited, average time spent on a product/category page, etc.
- *Anonymized user information*: information about the current user, including pages visited so far and time spent on each, and content of the shopping basket. If the user can be identified (i.e., is logged in) also historical data for this person, such as previous purchases and favourited products.
- *Trade log*: number of purchases/pieces sold for a given product, crosssales, most popular products, etc.

This data API can be used to develop services for the various product search and recommendation tasks (i.e., **C**). Table 1 summarizes the possible interfaces for these tasks, where in the case of the product search task, the task API takes a query as input and returns a ranked list of product ids as output. Once such a service is developed, researchers could then invoke the evaluation forum's intrinsic evaluation, that calls on their services, to evaluate the given component.

B) **Commercially Available Application.** The data from the live applications is supplied via this web service. This requires an online shop to participate in the living lab, supply the usage data and to trial research methods.

C) **Research Components as Web Services.** This web service defines a task API for each of the search and recommendation tasks addressed (i.e., Table 1). The developed methods then can be tested either using intrinsic evaluation (by using web services of **A**) or within the living lab application (utilized by services of **B**). Intrinsic evaluation allows for the component to be compared against other methods, as well as tested against any judgements acquired for the task from sandbox evaluation performed with assessors. However, evaluation within the living lab makes it possible to measure the user experience of customers over entire sessions, quantified e.g., in terms of time spent on the site, conversion rate or the sum of purchases made.

D) Non-Commercial Variants of Live Applications. The first variant is the Test Center Application, which allows users to observe and examine users' interactions with the live online shop system through usage logs collected. The second variant is the Sandbox Application; it is the implementation of a given task submitted by a researcher to be evaluated in the live system.

We have described the high level interactions between services for an online shop based living lab. Realization of this vision will require a substantive amount of negotiation between evaluation organizers, an online retailer and researchers to come to agreement about what can and can not be accommodated. The specific details about what data can be provided by commercial organizations will invariably determine what tasks can be acceptably outsourced to external researchers and under what conditions. If the conditions and restrictions are too great, then an alternative solution may involved setting up a dedicated commercial web application for research purposes. Assuming it is possible to amicably involve a commercial web application in the process, decisions about what search tasks and measures can be undertaken to define the APIs and types of intrinsic evaluations.

Here, we have only covered the high-level aspects regarding the design and development of a living lab for online shopping. While, still quite abstract, we hope this leads to some meaningful dialogue within the community and facilitates the development of a living lab for product search evaluation.

5 Discussion and Future Plans

In this paper we have outlined a potential architecture for developing the infrastructure to support a living lab in the context of IR evaluation. To provide a concrete example and propose a new evaluation track, we discussed what a lab might look like in an online shopping environment, for product search and related tasks. Central to the design is a distributed and flexible web based architecture (i.e., service oriented), and this means that a number of parties can cooperate in an independent fashion. However, there are a number of issues that such an evaluation platform would need to address.

What are the problems and challenges that face the development and use of living labs? There are a number of legal and ethical issues that need to be considered (such as, user consent and ethics approval of such research, legalities regarding the release of data, copyright issues, commercial sensitivity of interaction data, trust between parties), as well as privacy and security issues for the users and the commercial organizations (think AOL query log fiasco). A concern that may put off commercial organizations releasing data is the (perceived) commercial value of the said data and exposing part of their business processes. This may lead to competitors gaining an advantage. Legal, ethical and business issues aside, there are also a number of technical challenges which arise and range from design and implementation issues, to the cost of implementation, maintenance and adoption, to the reliability, robustness and provision of services. Here we have focused on the architecture to provide a possible design. Once the machinery is designed and created, the next barriers are in terms of its adoption and use by researchers and developers, and importantly the cooperation and support of the commercial organizations involved. Management issues will invariably arise in the how the evaluation forum and infrastructure is managed, and who should be responsible for maintaining such as service. To resolve these issues will either require a dedicated group of volunteers and/or long-term funding to maintain, organize and coordinate services. However, it may be feasible to prototype living lab on a small budget, if it was run for a short duration and a limited number of participants.

What are the benefits of a developing a living lab? One of the key benefits for researchers would be the access to real interaction data and (a variety of) real application contexts (like product search). Evaluations would become more user focused, and enable many more tasks to be evaluated and explored. The methods developed by researchers would have the potential to improve business processes. Thus labs lend themselves to being a bridge between academia and industry providing a direct route to commercialization for researchers. Besides access to more data and commercialization, another benefit of a living lab is that it can facilitate the independent verification of research results. This is because the evaluation forum services and commercial organizations can validate the research independently. Commercial organizations that participate in such initiatives could also benefit from having access to research and development teams without the associated overheads. Improvements to the provision of their service could lead to substantial improvements to their bottom line. In particular, for smaller organizations (such as independent online shops) and non-profit organizations (such as ACM Portal, citeseer.com, etc.) that cannot afford research staff, participation means having access to expertise with minimal investment. Also, participation would enable organizations to perform controlled experiments with good return-on-investment [10]. With appropriate infrastructure that facilitates experimentation and evaluation organizations could also innovate faster and more effectively [10].

Outlook and future directions. These are only some of the challenges, issues and benefits regarding the creation and development of a living lab. The major problems that need to be overcome are: (1) the initial design and development of the infrastructure to support a lab, and (2) the commitment of an organization and access to their data. While, the costs of building and developing the infrastructure are likely to be quite high for a fully integrated living lab, it may be possible to create a light-weight or scaled-down version on a smaller budget. Secondly, having an organization agree and commit to providing the tasks and data to support those tasks being performed is required. This is where we believe focusing on smaller online vendors or services would be more successful, than trying to develop living lab for a web search engine. Smaller vendors have specific problems and rich interaction data and are often without the resources to invest heavily in research. To this end, we are currently discussing with small online retailers about participating in such an initiative. However, before we continue to develop this initiative further we would like to discuss the proposed living lab on product search with the wider community; ascertain the level of interest, the potential concerns and inevitable constraints, as well as discuss the possibility of developing and organizing a product search evaluation campaign as part of a forum such as CLEF.

References

[1] Allan, J.: Hard track overview in trec 2003: High accuracy retrieval from documents. In: Text REtrieval Conference (2003)
[2] Azzopardi, L., Järvelin, K., Kamps, J., Smucker, M.D.: SIGIR 2010 workshop on the simulation of interaction. SIGIR Forum 44, 35–47 (2011)
[3] Belkin, N.J.: Some(what) grand challenges for information retrieval. SIGIR Forum 42(1), 47–54 (2008)
[4] Broder, A., Ciaramita, M., Fontoura, M., Gabrilovich, E., Josifovski, V., Metzler, D., Murdock, V., Plachouras, V.: To swing or not to swing: learning when (not) to advertise. In: Proc. of the 17th ACM Conf. on Information and Knowledge Management, CIKM 2008, pp. 1003–1012 (2008)
[5] Dumais, S.T., Belkin, N.J.: The trec interactive tracks: Putting the user into search. In: Text REtrieval Conference (2005)
[6] Jarvenpaa, S.L., Todd, P.A.: Consumer reactions to electronic shopping on the world wide web. Int. J. Electron. Commerce 1, 59–88 (1996)
[7] Kamps, J., Geva, S., Peters, C., Sakai, T., Trotman, A., Voorhees, E.: Report on the SIGIR 2009 workshop on the future of IR evaluation. SIGIR Forum 43, 13–23 (2009)
[8] Kelly, D.: Methods for evaluating interactive information retrieval systems with users. Foundations and Trends in Information Retrieval 3, 1–224 (2009)
[9] Kelly, D., Dumais, S., Pedersen, J.O.: Evaluation challenges and directions for info. seeking support systems. Computer 42(3), 60–66 (2009)
[10] Kohavi, R., Longbotham, R., Sommerfield, D., Henne, R.M.: Controlled experiments on the web: survey and practical guide. Data Min. Knowl. Discov. 18, 140–181 (2009)
[11] Pirolli, P.: Powers of 10: Modeling complex information-seeking systems at multiple scales. Computer 42, 33–40 (2009)
[12] Sanderson, M.: Test collection based evaluation of information retrieval systems. Foundations and Trends in Information Retrieval (FnTIR) 4(4), 247–375 (2010)
[13] Sparck-Jones, K.: Information Retrieval Experiment. Butterworth & Co. (1981)
[14] Voorhees, E.M., Harman, D.K.: TREC: Experiment and Evaluation in Information Retrieval. MIT Press, Cambridge (2005)
[15] White, R.W., Muresan, G., Marchionini, G.: SIGIR 2006 Workshop on Evaluating Exploratory Search Systems. SIGIR Forum 40, 52–60 (2006)

A Comparison of Evaluation Metrics for Document Filtering⋆

Enrique Amigó, Julio Gonzalo, and Felisa Verdejo

UNED NLP & IR Group
Juan del Rosal, 16
28040 Madrid, Spain
http://nlp.uned.es

Abstract. Although document filtering is simple to define, there is a
wide range of different evaluation measures that have been proposed in
the literature, all of which have been subject to criticism. We present a
unified, comparative view of the strenghts and weaknesses of proposed
measures based on two formal constraints (which should be satisfied by
any suitable evaluation measure) and various properties (which help differentiating measures according to their behaviour). We conclude that
(i) some smoothing process is necessary process to satisfy the basic constraints; and (ii) metrics can be grouped into three families, each satisfying one out of three formal properties, which are mutually exclusive,
i.e. no metric can satisfy all three properties simultaneously.

1 Introduction

Document Filtering is a generic problem that includes a wide set of tasks such
as spam detection [5], information retrieval over user profiles [10] or blog post
retrieval for on-line reputation management [1]. In essence, document filtering is
a binary classification task which consists of selecting relevant documents from
an input stream.

Although document filtering is simple to define, there is a wide range of different evaluation metrics that have been proposed in the literature, all of which have
been subject to criticism. Finding an optimal evaluation measure is, indeed, a
challenging problem. First, trivial non-informative baselines (e.g. filtering everything or discarding everything) might have a competitive performance depending
on the nature of the corpus (e.g. the average rate of relevant documents) and how
we evaluate performance. Second, systems should be evaluated over test cases -
topics - with variable ratios of relevant documents (for instance, every company
name has its own degree of ambiguity in search results for online reputation management: results for Apple are much noisier than results for Microsoft), and an
appropriate measure should be robust to differences between stream test cases.

⋆ This research was partially supported by the Spanish Ministry of Science and Innovation (Holopedia Project, TIN2010-21128-C02) and the Regional Government of
Madrid and the European Social Fund under MA2VICMR (S2009/TIC-1542) .

P. Forner et al. (Eds.): CLEF 2011, LNCS 6941, pp. 38–49, 2011.

Third, the penalization for negative samples in the output should be adjustable to the task and usage scenario: for instance, a spam e-mail or a non relevant blog post imply different processing efforts for the user which should be considered by the measure. All these issues make the evaluation problem non trivial.

Our goal is to acquire a unified view of the criticisms that have been made about the different measures used for the problem; to achieve this goal, we define two formal constraints (that any metric should satisfy) and three additional metric properties (that help understanding differences between current metrics). This leads to a comparative analysis of the strenghts and weaknesses of current metrics and how they complement each other. According to our analysis, we conclude that (i) metrics can be grouped into three families, each satisfying one of our three formal properties, which are mutually exclusive: no metric can satisfy all three properties simultaneously; and (ii) all metrics can be tuned (slightly redefined) in order to satisfy the basic constraints.

2 State of the Art

Typically, a filtering system – and any binary classification system in general – outputs a probability of relevance for every item[1], and the final classification implies choosing a threshold for this probability. Then, items above/below the threshold are classified as relevant/irrelevant. One way of evaluating document filtering is by inspecting the rank of documents (ordered by decreasing probability of relevance) and then measuring precision and recall at certain points in the rank. The advantage of this type of evaluation is that the classification algorithm can be evaluated independently from how the threshold is finally set.

For classification problems in general, an example of this type of evaluation is ROC (Receiver Operating Characteristic) [17], which computes the area under the precision/recall curve for decreasing threshold values, or AUC [14]. For document filtering tasks, some researchers evaluate Precision at a certain number of retrieved documents [19,3] or average across recall levels [16] – i.e., precision at a fixed rank cutoff–. Other related measures are the Mean Cross-entropy [8], Root-mean-squared error, Calibration Error [7], SAR and Expected Cost (all of them available and described, for instance, in the ROCR package of the R tool[2]).

This type of measures has been criticised [20] because they do not consider the ability of systems to predict the ratio of relevant documents in single streams, which is a crucial aspect of system quality. In other words, metrics should work on the binary output of the filtering system rather than on the internal rank produced by the system to make the final binary decisions. In this paper we will focus on measures that assume a binary system output.

Most classification measures assume a binary system output. First, we have to discard partial measures; i.e., those for which a maximum score does not imply necessarily a perfect output, such as False Positive Rate, False Negative Rate,

---

[1] Or, more precisely, a quantity which can be mapped into a probability of relevance using some growing monotonic function.

[2] http://cran.r-project.org/web/packages/ROCR/ROCR.pdf

Recall, True Negative Rate, Precision, Negative Predictive Value, Prediction-conditioned Fallout, Prediction-conditions Miss, Rate of Positive Predictions or Rate of Negative Predictions. The common aspect of these measures is that they only consider two of the four components of the contingency matrix. Knowing the total number of samples, at least three components are necessary to ensure the perfect quality of maximum scored outputs.

We now focus on non-partial measures employed in filtering tasks that assume a binary system output. In [11], Precision and Recall are combined by means of the product. The standard F measure [22] has also been used in document filtering evaluation campaigns [9]. In general, the main criticism against Precision/Recall based measures is that, when no relevant documents are returned, a system receives a score of zero, regardless of the system output size.

Accuracy (fraction of correctly classified items) has been employed in some cases for filtering tasks [6,23,1]. As for any binary classification in general, the accuracy measure is adequate only when relevant and irrelevant documents are well balanced in the input stream. In addition, it does not include a weighting parameter to establish the relative cost of missing relevant documents versus returning irrelevant documents. Because of this, some authors prefer Utility-based measures. This family of measures consists of a linear combination of the four components in the contingency table (true positives, true negatives, false positives and false negatives) The most common Utility measure is a linear combination of true positives and false positives, i.e. all retrieved items [11,12,10,19]. Given that Utility does not have a defined range, it is usually scaled [12,10]: The main criticism about Utility based measures is well described in [11]: "scores will vary widely from topic to topic based on the number of relevant documents, and there is no valid way to normalize them, meaning that the scores can not easily be averaged or compared across topics".

In certain filtering tasks, the *logistic average misclassification percentage* (Lam%) is the preferred evaluation measure. For instance, in the Spam Filtering Track of TREC evaluation campaigns [5] In [18] an important drawback of *lam%* is detected. When either the missclassified relevant or non relevant documents is zero, *lam%* is maximal regardless of the other partial measure. The practical implication is that a system with a strong classification threshold (i.e. predicting very few positive cases) can achieve a top score without predicting the correct class in most cases.

Other measures that have not been employed in filtering tasks are the Phi correlation coefficient, the odds ratio and the chi square test statistics. In this paper, we will see that these measures are closely related with some of the previous measures.

Overall, all present measures are perceived to have drawbacks, and consequently there is no consensus yet about what is the most appropriate evaluation measure for filtering tasks. There is more consensus, however, on the desired properties that an optimal metric should satisfy. We will start by defining such properties as formal constraints.

3 Task Definition and Formal Constraints

For the sake of readability, we will focus on document filtering. However, the conclusions in this paper can be applied to any other filtering task. We assume that the filtering system must return a binary classification (rather than a document ranking). According to this, the most simple way of representing filtering outputs is as a subset $S \subset T$ of the input stream T. S is the subset of documents which the system consider relevant, and is evaluated against a gold standard G (the subset of documents which are truly relevant). A metric returns a certain score $Q(S)$ for the output set S. We will use the notation: $P(S) = P(e \in S)$ and $P(G|S) \equiv p(e \in G|e \in S) \sim \frac{|S \cap G|}{|S|}$.

The document filtering task has some particularities with respect to other classification tasks: (i) Systems must be evaluated over a certain number of independent input streams ("topics"). For instance, user accounts for spam filtering, user profiles for personalized retrieval, or company names for on-line reputation management. In addition, the ratio of positive samples (relevant documents) across topics can be highly variable; (ii) The penalization for negative samples (non-relevant documents) in the output can be subjective. That is, a spam e-mail or a non relevant post in the output stream implies a certain effort for the user that depends on the task and; (iii) The relative penalization for errors (false negatives versus false positives) depends on each particular scenario, and this relative cost should be taken into account for evaluation.

We now start by defining two basic formal constraints which should be satisfied by any suitable evaluation metric. The first one is the *Best System Constraint*. The system quality is maximum if and only if the system output matches exactly the gold standard:

$$S = G \longrightarrow \forall S'.Q(S) \geq Q(S')$$

The second one is the *Growing Quality Constraint*. Given a certain system output, if we consider a variation of the output in which an element that was misclassified is now correctly classified, the quality of the transformed output must be higher than the original one. Equivalently: adding an irrelevant document to the output S must decrease the score, and adding a relevant document to S must increase the score. Using set notation:

$$Q(S \cup \{e \in \neg G\}) < Q(S) \quad \text{and} \quad Q(S \cup \{e \in G\}) > Q(S)$$

4 Measure Properties

Systems can achieve very different results depending on the evaluation measure employed; however, the consequences of choosing a given measure have not yet been clarified in the state of the art. Our goal is to define a set of properties that help explaining the comparative behavior of measures, and therefore choosing the the most appropriate for each case.

We claim that the main difference between metrics is how a *non informative* system is evaluated. Non informative systems are those in which the output set S is independent of the input:

$$P(S|G) = P(S) \quad \wedge \quad P(G|S) = P(G)$$

For instance, any random selection of n elements from the input is a non-informative solution; the "all true" baseline is a special case with n being the number of items in the input; and the "all false" is the special case where $n = 0$. The question is how evaluation measures handle non-informative systems.

Absolute Weighting Property. Consider the following scenario: a phone company wants to predict which clients are about to leave the company from a number of signals (for instance, phone conversations with former clients that have just left the company). The prediction is a set of clients, which will then be offered a reduction in their bills to avoid losing them. For every component of the contingency matrix there is an absolute gain associated with each classified element: for instance, every true positive implies a gain which is the difference between keeping a client or losing it.

We define the *Absolute Weighting* property as the ability of measures to assign an absolute weight to relevant (versus non relevant) documents in the output regardless of the output size. We formalize this property as follows: there exists a parameter c in the evaluation measure that determines if removing one document from each class (relevant/not relevant) from the output S improves it. Formally, being:

$$S_2 \equiv S_1 - \{e_G \in G\} - \{e_{\neg G} \in \neg G\} \quad \text{then} \quad \exists th. (c > th \leftrightarrow Q_c(S_1) > Q_c(S_2))$$

Non Informativeness Fixed Quality Property. In other scenarios, non-informative outputs are equally useless. For instance, suppose that the user needs a directory of high-priority emails which should be read first. In this situation, a non-informative system will always result in a similar proportion of true positives in both directories (high-priority vs. others), which is useless.

We formalize the *Non Informativeness Fixed Quality* property as follows:

$$p(G|S) = p(G) \longrightarrow Q(S) = k$$

where k is a constant.

Obviously, the *Absolute Weighting* and the *Non Informativeness Fixed Quality* properties are mutually exclusive.

Non Informativeness Decreasing Quality Property. Now, let us consider a spam filtering task. Here the goal consists of removing spam from an email stream. A non informative system would remove e-mails regardless of their content. In these conditions, filtering nothing is better than removing e-mails randomly. What we need here is a *Non Informativeness Decreasing Quality* property: The more a non informative output reduces the input stream, the more its score is reduced.

Table 1. Basic constraints, properties and measures

	Basic Constraints		Properties		
	Best System	Growing Quality	Absolute Weighting	Non Inform. Fixed Quality	Non Inform. Decreasing Quality
Acc	YES	YES	NO	NO	NO
Weighted Acc	YES	YES	YES	NO	NO
Utility	YES	YES	YES	NO	NO
Norm. Utility	YES	NO	YES	NO	NO
Lam%	NO	NO	NO	YES	NO
Lam%$_{smooth}$	YES	YES	NO	YES	NO
Phi, MAAC, Kaps	YES	YES	NO	YES	NO
Odds, MI, Chi test	YES	NO	NO	YES	NO
F Measure	YES	NO	NO	NO	YES
F$_{smooth}$	YES	YES	NO	NO	YES

$$p(G|S) = p(G) \longrightarrow Q(S) \sim |S|$$

Again, this property is incompatible with the previous two. Here, non-informative systems get different scores (violating *Non Informativeness Fixed Quality*). In addition, an absolute weighting for true and false positives (*Absolute Weighting* Property) can break the correlation between the score and the size of non informative outputs.

According to these properties we can distinguish between three metric families, which are analyzed in next section.

5 Utility-Based Metrics

Utility based metrics are those that can be expressed as a linear combination of the four components in the contingency matrix [12]: true positives ($|S \cap G|$), true negatives ($|\neg S \cap \neg G|$), false positives ($|S \cap \neg G|$) and false negatives ($|\neg S \cap \neg G|$). Usually, the resulting score is scaled according to the size of the input stream used for evaluation ($|G|$ and $|\neg G|$).

The Accuracy measure (proportion of correctly classified documents) and the error rate (1-Accuracy) are two particular cases of Utility measures which reward equally $|S \cap G|$ and $|\neg S \cap \neg G|$. The result is scaled over the input stream size. Implicitly, accuracy penalizes also the $|\neg S \cap G|$ and $|S \cap \neg G|$ components. The Accuracy measure can be expressed as $\frac{|S \cap G| + |\neg S \cap \neg G|}{|T|}$.

In [2] a weighted version of Accuracy is proposed: $\frac{\lambda |S \cap G| + |\neg S \cap \neg G|}{\lambda |G| + |\neg G|}$ Basically, the Weighted Accuracy is an Utility measure that assigns a relative weight to $|S \cap G|$ and normalizes the score according to the function $\frac{1}{\lambda |G| + |\neg G|}$.

The most common Utility version assigns a relative α weight between true positives and false positives:

$$U(S) = \alpha |S \cap G| - |S \cap \neg G|$$

This Utility version is normalized by means of a scaling function:

$$U_s^*(S,T) = \frac{max(U(S,T), U(s)) - U(s)}{MaxU(T) - U(s)}$$

where $u(S,T)$ is the original utility of the system output S for topic T, $U(s)$ is the utility of retrieving s non-relevant documents, and $MaxU(T)$ is the maximum possible utility score for topic T.

Accuracy satisfies both the *Best System* and the *Growing Quality* constraints. As for the (most popular) normalized Utility metric, satisfies the *Best System* constraint and also the *Growing Quality* constraint, except when applying the scaling function, where the term $max(u(S,T), U(s))$ violates the constraint for the cases where $u(S,T) \leq U(s)$.

The most characteristic property of Utility-based metrics is that they satisfy the *Absolute Weighting* property. Being:

$$S_2 = S_1 - \{e_G \in G\} - \{e_{\neg G} \in \neg G\} \quad \text{then} \quad \exists c.(h > c \leftrightarrow Q(S_1) > Q(S_2))$$

In the case of Utility, the c value is 1:

$$Utility(S_2) = \alpha|S_2 \cap G| - |S_2 \cap \neg G| = \alpha(|S_1 \cap G| - 1) - (|S_1 \cap \neg G| - 1) =$$

$$= Utility(S_1) - \alpha + 1 > Utility(S_1) \text{ if } \alpha < 1$$

Although Accuracy can be considered an Utility-based measure, it does not directly satisfy the Absolute Weighting Property, given that its definition does not include any parameter. However, the weighted accuracy proposed in [2] also satisfies this property.

6   Informativeness-Based Measures

This family of measures satisfy the second property *Non Informativeness Fixed Quality*. That is, it includes measures that scores equally any non-informative solution, We first focus on Lam% which is the most popular metric within this family.

Lam% was defined for the problem of spam detection as the geometric mean of the odds of $hm\%$ (ratio of misclassified ham) and $sm\%$ (ratio of misclassified spam). The Lam% scale is reversed; maximum Lam% represents minimum quality:

$$lam\% = logit^{-1}(\frac{logit(hm\%) + logit(sm\%)}{2})$$

$$hm\% = \frac{|\neg S \cap G|}{|G|} \quad sm\% = \frac{|S \cap \neg G|)}{|\neg G|} \quad logit(x) = log(\frac{x}{1-x})$$

The main criticism to Lam% is that when either $hm\%$ or $sm\%$ are zero, $lam\%$ is minimal (maximal quality) regardless of the other value [18]. This phenomenon prevents both the *Best System* and the *Growing Quality* constraints from being satisfied. But this is an effect of data granularity and how misclassification is estimated, rather than an intrinsic drawback of Lam%. The solution consists of applying some kind of smoothing, such that $hm\%$ and $sm\%$ can not be zero.

The most characteristic property of Lam% is that it assigns a fixed score to every non-informative system ($Lam\%(S) = 1/2$), regardless of the ratio of relevant documents in the input stream (*Non Informativeness Fixed Quality* property). Remember that non-informative approaches are those that randomly select a percentage of the documents in the input stream. Therefore, the ratio of relevant documents in the output is the same as in the input stream: $p(G|S) = p(G)$.

Let us prove that Lam% satisfies this property. Given a non informative output S', then:

$$hm\%(S') = \frac{|\neg S' \cap G|}{|G|} = p(\neg S'|G) = \frac{p(G|\neg S')p(\neg S')}{p(G)} = \frac{p(G)p(\neg S')}{p(G)} = \frac{|\neg S'|}{|T|}$$

$$sm\%(S') = \frac{|S' \cap \neg G|)}{|\neg G|} = p(S'|\neg G) = \frac{p(\neg G|S')p(S')}{p(\neg G)} = \frac{p(\neg G)p(S')}{p(\neg G)} = p(S') = \frac{|S'|}{|T|}$$

The resulting lam% score is

$$lam\%(S') = logit^{-1}(\frac{logit(sm\%(S')) + logit(hm\%(S'))}{2})$$

We can show that when the system is not informative, the two expressions in the numerator cancel each other:

$$logit(sm\%(S')) = logit(\frac{|\neg S'|}{|T|}) = log(\frac{\frac{|\neg S'|}{|T|}}{1 - \frac{|\neg S'|}{|T|}}) = log\frac{|\neg S'|}{|T|} - log(1 - \frac{|\neg S'|}{|T|}) =$$

$$log(1 - \frac{|S'|}{|T|}) - log(\frac{|S'|}{|T|}) = -log\frac{|S'|}{|T|} - log(1 - \frac{|S'|}{|T|})) = -log\frac{\frac{|S'|}{|T|}}{1 - \frac{|S'|}{|T|}} = -logtit(hm\%(S'))$$

Therefore, given any non informative output S':

$$lam\%(S') = logit^{-1}(0) = \frac{e^0}{1 + e^0} = 1/2$$

On the other hand, if we include the weighting parameter, we can satisfy the *Absolute Weighting* property of Utility metrics. Removing non-relevant documents affects sm%, while removing relevant documents from the output affects hm%. However, the equality $logit(\alpha sm\%(S')) = -logit(hm\%(S'))$ for non informative outputs S' would no longer be satisfied and the *Non Informativeness Fixed Quality* property would be violated.

In summary, the main property of Lam% is that the score for any non-informative system is fixed, but this property is not compatible with the Utility metric property of penalizing noise in absolute terms.

There exist other measures that satisfy the *Non Informativeness Fixed Quality* property. Because of space availability, we do not include here a formal proof. These measures are: Phi correlation coefficient, The Macro Average Accuracy [15], the Kappa statistic [4], Chi square test statistic Odds ratio [13], and the Mutual Information (MI).

7   Class-Oriented Measures: Precision and Recall

The third measure family includes those that assume a certain asymmetry between classes. Precision/recall-based measures are suitable for applications where one data class is of more interest than others such as, for instance, Information Retrieval tasks[21].

The most representative measure in this family is the combination of precision and recall. We will focus here on the standard F measure combining function [22], but the same conclusions can be applied to the P & R product [11]. We will use the notation $\Re$ for Recall and $\wp$ for Precision in order to avoid confusion with probabilities.

$$F_\alpha(S) = \frac{1}{\frac{\alpha}{\wp(S)} + \frac{1-\alpha}{\Re(S)}} \quad \text{where} \quad \wp(S) = \frac{|S \cap G|}{|S|} \quad \Re(S) = \frac{|S \cap G|}{|G|}$$

We can express Precision and Recall in terms of probabilities ($\wp(S) = p(G|S)$ and $\Re(S) = p(S|G)$).

The F measure over Precision and Recall satisfies the *Best System Constraint* given that just the gold standard solution achieves a maximum Precision and Recall.

$$\wp(S) = 1 \wedge \Re(S) = 1 \leftrightarrow |S \cap G| = |S| \wedge |S \cap G| = |G| \leftrightarrow |S| = |G|$$

The main criticism in the state of the art is that Precision is not able to distinguish between outputs that contain only irrelevant documents. This affects the *Growing Quality* constraint. Precision is zero for any output without relevant documents. Therefore, according to the *Decreasing Marginal Effectiveness* property of F [22], the score is always zero regardless of the amount of irrelevant documents in the output. However, just like in the case of Lam%, this is not an intrinsic drawback of the metric, but a problem of how precision is estimated. It can be solved by applying a smoothing method.

The F measure (and its metric family) is distinguishable from other measures in the way that it scores non-informative outputs (i.e. when $p(G|S) = p(G)$): the more a non informative output reduces the input stream, the more its F value is reduced. Thus, they satisfy the *Non Informativeness Decreasing Quality* property:

$$(p(G|S) = p(G) \longrightarrow Q(S) \sim |S|$$

Let us prove it: Precision for a non-informative output S is fixed given a certain input stream T, and Recall always grows with the output size of the non-informative system:

$$\wp(S') = p(G|S) = p(G) \quad \wedge \quad \Re(S) = p(S|G) = \frac{p(G|S)p(S)}{p(G)} = \frac{p(G)p(S)}{p(G)} = \frac{|S'|}{|T|}$$

Therefore, the non-informative output score is correlated with its size. In other words, reducing randomly the input stream decrements F.

As we described in Section 4, this property is not compatible with the *Non Informativeness Fixed Quality* property satisfied by Lam%. It is also incompatible with the *Absolute Weighting* property. An absolute penalization can produce a growing quality when filtering randomly the input stream. More concretely, when adding a relevant document and removing an irrelevant document from the output, Recall increases. On the other hand, Precision also increases if it was lower than 0.5:

$$P(S - \{e \in \neg G\} - \{e \in \neg G\}) = \frac{|S \cap G| - 1}{|S| - 2}$$

$$\frac{|S \cap G| - 1}{|S| - 2} > \frac{|S \cap G|}{|S|} \leftrightarrow \frac{|S \cap G|}{|S|} < 0.5$$

Therefore, both Precision and Recall increase. Due to the F measure *Independence* property [22], F also grows for any α value and the *Absolute Weighting* property is not satisfied.

8 Conclusions

The current variety of approaches to document filtering evaluation seems more a reflection of the lack of a systematic comparison of the properties of each metric than a consequence of the different nature of the various filtering tasks. Just as an illustration, TREC has organized at least three filtering tasks, all of them using different evaluation metrics: the filtering track used utility [12], the spam track chose Lam% [5], and the legal track has employed a variation of F [9]. On the other hand, the WePS-3 online reputation management task [1], which is also a filtering task, uses accuracy as the official ranking measure.

Our work attempts to fill this gap by presenting a comparison of measures based on formal constraints and properties. Our analysis shows that evaluation measures for document filtering can be grouped into three families, each satisfying one out of three formal properties, which are mutually exclusive, i.e. no metric can satisfy all three properties simultaneously. It also shows that all metrics can be adapted to satisfy two basic constraints which we propose as formal constraints on any suitable document filtering measure.

We are not prescribing any of the standard measures as the best option for every conceivable document filtering scenario. Only our two basic constraints are formal requisites for a valid evaluation measure, and we have shown that most current metrics can be adapted to satisfy them. The rest of the properties defined in our work are primarily a tool to understand the differences and similarities between measures, rather than requisites. Indeed, there may be certain scenarios in which, depending on the dataset and the type of systems and baselines considered, one or more properties may be innecesary or even considered harmful.

References

1. Amigó, E., Artiles, J., Gonzalo, J., Spina, D., Liu, B., Corujo, A.: WePS3 Evaluation Campaign: Overview of the On-line Reputation Management Task. In: 2nd Web People Search Evaluation Workshop (WePS 2010), CLEF 2010 Conference, Padova Italy (2010)
2. Androutsopoulos, I., Koutsias, J., Chandrinos, K., Paliouras, G., Spyropoulos, C.D.: An evaluation of naive bayesian anti-spam filtering. CoRR cs.CL/0006013 (2000)
3. Callan, J.: Document filtering with inference networks. In: Proceedings of the Nineteenth Annual International ACM SIGIR Conference on Research and Development in Information Retrieval, pp. 262–269 (1996)
4. Cohen, J.: A Coefficient of Agreement for Nominal Scales. Educational and Psychological Measurement 20(1), 37 (1960)
5. Cormack, G., Lynam, T.: Trec 2005 spam track overview. In: Proceedings of the fourteenth Text Retrieval Conference 8TREC 2005 (2005)
6. Cunningham, P., Nowlan, N., Delany, S.J., Haahr, M.: A case-based approach to spam filtering that can track concept drift. In: The ICCBR 2003 Workshop on Long-Lived CBR Systems, pp. 03–2003 (2003)
7. Fawcett, T., Niculescu-Mizil, A.: Pav and the roc convex hull. Mach. Learn. 68, 97–106 (2007)
8. Good, I.J.: ational decisions. Journal of the Royal Statistical Society. Series B Methodological 14, 107–114 (1952)
9. Hedin, B., Tomlinson, S., Baron, J.R., Oard, D.W.: Overview of the trec 2009 legal track (2009)
10. Hoashi, K., Matsumoto, K., Inoue, N., Hashimoto, K.: Document filtering method using non-relevant information profile. In: Proceedings of the 23rd Annual International ACM SIGIR Conference on Research and Development in Information Retrieval, SIGIR 2000, pp. 176–183. ACM, New York (2000),
http://doi.acm.org/10.1145/345508.345573
11. Hull, D.A.: The trec-6 filtering track: Description and analysis. In: Proceedings of the TREC 6, pp. 33–56 (1997)
12. Hull, D.A.: The TREC-7 filtering track: description and analysis. In: Voorhees, E.M., Harman, D.K. (eds.) Proceedings of TREC-7, 7th Text Retrieval Conference, pp. 33–56. National Institute of Standards and Technology, Gaithersburg (1998),
citeseer.ist.psu.edu/126480.html
13. Karon, B.P., Alexander, I.E.: Association and estimation in contingency tables. Journal of the American Statistical Association 23(2), 1–28 (1958),
http://www.jstor.org/stable/2283825
14. Ling, C.X., Huang, J., Zhang, H.: Auc: a statistically consistent and more discriminating measure than accuracy. In: IJCAI, pp. 519–526 (2003)
15. Mitchell, T.M.: Machine learning. McGraw Hill, New York (1997)
16. Persin, M.: Document filtering for fast ranking. In: Proceedings of the 17th Annual International ACM SIGIR Conference on Research and Development in Information Retrieval, SIGIR 1994, pp. 339–348. Springer, New York (1994),
http://portal.acm.org/citation.cfm?id=188490.188597
17. Provost, F.J., Fawcett, T.: Analysis and visualization of classifier performance: Comparison under imprecise class and cost distributions. In: Knowledge Discovery and Data Mining, pp. 43–48 (1997)

18. Qi, H., Yang, M., He, X., Li, S.: Re-examination on lam% in spam filtering. In: Proceedings of the SIGIR 2010 Conference, Geneva, Switzerland (2010)
19. Robertson, S., Hull, D.A.: The trec-9 filtering track final report. In: Proceedings of TREC-9, pp. 25–40 (2001)
20. Schapire, R.E., Singer, Y., Singhal, A.: Boosting and rocchio applied to text filtering. In: Proceedings of ACM SIGIR, pp. 215–223. ACM Press, New York (1998)
21. Sokolova, M.V., Japkowicz, N., Szpakowicz, S.: Beyond accuracy, F-score and ROC: A family of discriminant measures for performance evaluation. In: Sattar, A., Kang, B.-h. (eds.) AI 2006. LNCS (LNAI), vol. 4304, pp. 1015–1021. Springer, Heidelberg (2006)
22. Van Rijsbergen, C.: Foundation of evaluation. Journal of Documentation 30(4), 365–373 (1974)
23. Wei, C.P., Chen, H.C., Cheng, T.H.: Effective spam filtering: A single-class learning and ensemble approach. Decis. Support Syst. 45(3), 491–503 (2008)

Filter Keywords and Majority Class Strategies for Company Name Disambiguation in Twitter*

Damiano Spina, Enrique Amigó, and Julio Gonzalo

UNED NLP & IR Group
Juan del Rosal, 16
28040 Madrid, Spain
{damiano,enrique,julio}@lsi.uned.es
http://nlp.uned.es

Abstract. Monitoring the online reputation of a company starts by retrieving all (fresh) information where the company is mentioned; and a major problem in this context is that company names are often ambiguous (*apple* may refer to the company, the fruit, the singer, etc.). The problem is particularly hard in microblogging, where there is little context to disambiguate: this was the task addressed in the WePS-3 CLEF lab exercise in 2010. This paper introduces a novel *fingerprint* representation technique to visualize and compare system results for the task. We apply this technique to the systems that originally participated in WePS-3, and then we use it to explore the usefulness of *filter keywords* (those whose presence in a tweet reliably signals either the positive or the negative class) and finding the majority class (whether positive or negative tweets are predominant for a given company name in a tweet stream) as signals that contribute to address the problem. Our study shows that both are key signals to solve the task, and we also find that, remarkably, the vocabulary associated to a company in the Web does not seem to match the vocabulary used in Twitter streams: even a manual extraction of filter keywords from web pages has substantially lower recall than an oracle selection of the best terms from the Twitter stream.

1 Introduction

Monitoring the online reputation of a company starts by retrieving all (fresh) information where the company is mentioned; and a major problem in this context is that company names are often ambiguous. For instance, the query "Apple" retrieves information about Steve Jobs' company, but also about apples (the fruit) Fiona Apple (the singer), etc. The problem becomes particularly hard in microblogging streams such as `Twitter.com`, because the available context to disambiguate each tweet is much smaller than in other media.

---

* This research was partially supported by the Spanish Ministry of Education via a doctoral grant to the first author (AP2009-0507) and the Spanish Ministry of Science and Innovation (Holopedia Project, TIN2010-21128-C02).

P. Forner et al. (Eds.): CLEF 2011, LNCS 6941, pp. 50–61, 2011.

Our work deals with the challenge of providing an unsupervised solution for this problem, i.e. a system that accepts a company's name (plus a representative URL) as input and provides a binary classification of tweets (which mention the company's name) as related or non-related to the company. In practice, this could be a filtering component for services such as `SocialMention.com`, where the query "apple" produces aggregated figures of *strength, sentiment, passion, reach*, etc. for all sources mentioning apple, whether they refer to the company, the fruit, the singer, etc.

This research is based upon two intuitive observations:

(i) **Filter Keywords:** Manual annotation can be simplified by picking up special keywords (henceforth called *filter keywords*) that isolate positive or negative information. For instance, "ipod" is a positive filter keyword for apple tweets[1], because its presence is a highly reliable indicator that the tweet is about the apple company. Reversely, "crumble" is a negative filter keyword for apple, because it correlates with non-related tweets. The intuition is that automatic detection of such filter keywords can be a valuable information to provide an automatic solution to the problem.

(ii) **Majority Class:** The ratio of positive/negative tweets is extremely variable between company names, and does not follow a normal distribution; on the contrary, it seems to have a skewed distribution (at least when considering a short time frame): typically, either most tweets are about the company, or most tweets are unrelated to the company. Therefore, predicting which of the situations hold for a certain company name might be a valuable input for algorithmic solutions to the problem.

Our goal is to provide quantitative evidence supporting (or rejecting) our intuitions. We will use the WePS-3 Task 2 test collection[2] [1], which is, to our knowledge, the first dataset built explicitly to address this problem. This paper also introduces the *fingerprint* representation technique, a visualization of system results that is particularly useful to understand the behavior of systems in classification/filtering scenarios where the test cases have variable class skews.

In Section 2 we start by briefly discussing the state of the art, and introducing the *fingerprint* representation. In Section 3 we test our two hypotheses, and in Section 4 we discuss how to locate filter keywords automatically.

2   State of the Art

Disambiguation of company names is a necessary step in the monitorization of opinions about a company. However, this problem is not tackled explicitly in most research on the topic, where it is normally assumed that query terms are not ambiguous in the retrieval process. The disambiguation task has been explicitly addressed in the WePS-3 evaluation campaign [1]. In this section we revisit the results of that campaign using a novel *fingerprint* representation technique that helps understanding and comparing system results.

---

[1] We will refer to microblog entries as tweets – being Twitter the most prominent example of microblogging network – in the remainder of the article.

[2] `http://nlp.uned.es/weps/weps-3`

2.1 WePS-3: Test Collection and Systems

The Online Reputation Management task of the WePS-3 evaluation campaign consists of filtering Twitter posts containing a given company name depending of whether the post is actually related with the company or not. The WePS-3 ORM task dataset consists of 52 training and 47 test cases, each of them comprising a company name, its URL, and a set of tweets consisting of an average of 435 tweets manually annotated as related/non-related to the company. A total of five research groups participated in the campaign. The best two systems were LSIR [7] and ITC-UT [8].

The LSIR system builds a set of profiles for each company, made of keywords extracted from external resources such as WordNet or the company homepage, as well as a set of manually defined keywords for the company and the most frequent unrelated senses for the company name. These profiles are used to extract tweet-specific features that are added to other generic features that give information about the quality of the profiles to label the tweets as related or unrelated with an SVM classifier. The ITC-UT system is based on a two step classification. Firstly, it predicts the class of each query/company name according to the ratio of related tweets of each company name and secondly applies a different heuristic for each class, basically based on the PoS tagging and the named entity label of the company name.

The SINAI system [3] also uses a set of heuristic rules based on the occurrence of named entities both on the tweets and on external resources like Wikipedia, DBPedia and the company homepage. The UVA system [6] does not employ any resource related to the company, but uses features that related with the use of the language on the collection of tweets (URLs, hashtags, capital characters, punctuation, etc.). Finally, the KALMAR system [4] builds an initial model based on the terms extracted from the homepage to label a seed of tweets and then uses them in a bootstrapping process, computing the point-wise mutual information between the word and the target's label.

In the WePS exercise, accuracy (ratio of correctly classified tweets) was used to rank systems. The best overall system (LSIR) obtained 0.83, but including manually produced filter keywords. The best automatic system (ITC-UT) reaches an accuracy of 0.75 (note that 0.5 is the accuracy of a random classification), and includes a query classification step in order to predict the ratio of positive/negative tweets. These results motivate us to analyze in depth the potential contribution of filter keywords and majority class signals.

2.2 A *Fingerprint* Visualization to Compare Systems

One of our intuitions is that predicting which is the majority class for a given microblogging stream is a substantial step towards solving the problem. The reason is that the ratio of related tweets does not follow a normal distribution: in general, for a given company name and a (small) period of time, either most tweets refer to the company, or most tweets are unrelated. Even in the WePS-3 dataset, where organizers made an effort to include company names covering all the spectra of positive ratio values, the ratio is very low or very high for many companies.

In this context, average performance measures (accuracy or F) are not sufficiently informative of the system's behavior. That's our main motivation to propose a *fingerprint* representation technique to visualize system results. Figure 1 illustrates our method, which consists of displaying the accuracy of the system (vertical axis) for each company vs. the ratio of related (positive) tweets for the company (horizontal axis). Each dot in the graph represents one of the test cases (i.e. the accuracy of the system for the set of tweets containing the name of one of the companies in the dataset).

The advantage of this representation is that the three basic baselines (all true, all negative and random) are displayed as three fixed lines, independently of the dataset. The performance of the "all true" baseline classification corresponds exactly with the proportion of true cases, and therefore is the $y = x$ diagonal in the graph. The "all negative", correspondingly, is represented by the $y = 1 - x$ diagonal. And, finally, the random baseline is the horizontal line $y = 0.5$.

Note that, for averaged measures – such as accuracy or F –, the results of the baselines depend on the dataset: for instance, the "all true" baseline depends on the average number of true cases in the corpus used for evaluation. As our fingerprint representation is constant for that baselines, it is easier to identify when a system is having a baseline-like approach, and for which subset of the data.

Figure 1 shows the fingerprint representation of the best runs per participant. The first observation is that the "all true" and "all false" baseline lines seem to have a "magnetic" attraction on the dots; in other words, somehow most systems have a tendency to make a "winner takes all" decision on each of the test cases. The second observation is that the fingerprint representation makes a clear distinction between each system's behavior. SINAI is the system that most clearly tends to apply a winner takes all strategy, with a strong preference for the "all false" choice. KALMAR results cluster around the "all true" baseline, with deviations towards randomness. The two best systems, LSIR manual and ITC-UT, have a tendency towards the "all false" baseline when the true ratio is lower than 0.5, and towards the "all true" baseline when the true ratio is greater than 0.5. In the area around 0.5, LSIR is able to improve on both baseline strategies, with the dots spreading over a "glass of martini" shaped area. Finally, UVA is the only system which has no baseline behavior at all, with accuracy values which seem largely independent on the related tweets ratio.

The fingerprint representation of WePS-3 systems proves to be very useful to understand systems' behavior, and highlights the importance of the related tweets ratio: the best systems behave as if they were guessing which is the majority class for each test case, and most of the rest tend to behave like one of the baselines.

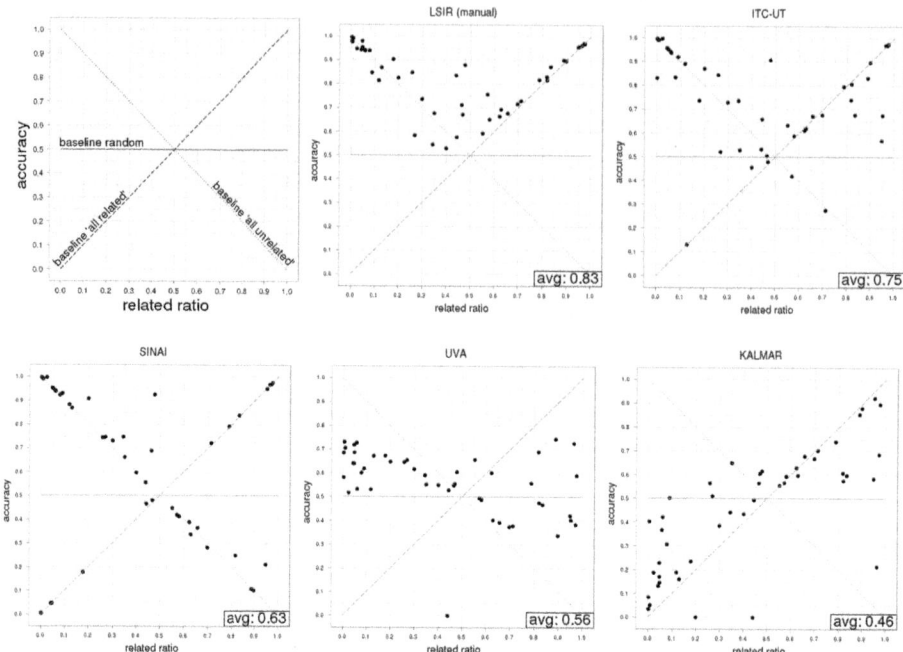

Fig. 1. Fingerprint technique and visualization of the best runs of participants in the WePS-3 exercise

3 Hypotheses Testing

3.1 Upper Bound Performance of Filter Keywords

A positive/negative filter keyword is an expression that, if present in a tweet, indicates a high probability that it is related/non-related to the company. The most useful filter keywords are those with a high recall, i.e., those which appear in as many tweets as possible. As all tweets in the WePS-3 collection are manually annotated as related/non-related to their respective company name, we can find exactly how many filter keywords there are (by definition, filter keywords are those terms that only appear in either the positive or the negative tweets), and how much recall they provide. Figure 2a shows the recall of the first n filter keywords (for $n = 1 \dots 20$) in the test collection. Recall at step n is the proportion of tweets covered by adding the keyword that filters more tweets among those which were not still covered by the first n-1 keywords. We will hereafter refer to these perfect keyword selection as *oracle keywords*.

The graph shows that, in average, the best five oracle keywords cover around 30% of the tweets, and the best ten reach around 40% of the tweets. That is, only the best five discriminative terms directly cover around 130 out of 435 tweets in each stream, which could in turn be used to build a supervised classifier

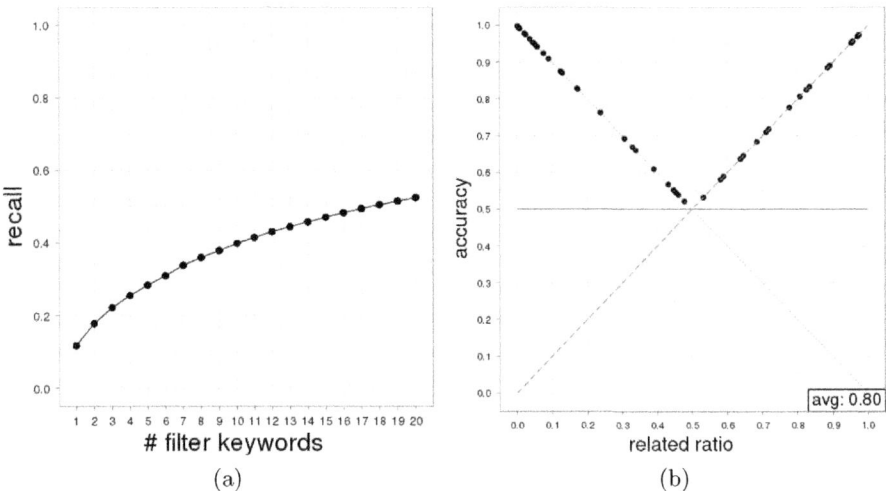

Fig. 2. Upper bounds of the two hypotheses: filter keywords (a) and majority class (b)

to be applied on the remaining tweets. This indicates that filter keywords are potentially a relevant source of information to address the problem.

A priori, the natural place to find filter keywords is the Web: the company's web domain, references to this domain in Wikipedia, ODP, etc. and the Web at large. Using the company's URL and web search results for the company name, a human annotator performed a manual selection of positive and negative keywords for all the companies in WePS-3 corpus. Note that the annotator did not have access to the tweets in the corpus.

Remarkably, manual keywords extracted from the Web (around 10 per company) only reach 14.61% coverage of the tweets (compare with 39.97% for 10 oracle keywords extracted from the tweets themselves), with an accuracy of 0.86 (lower than expected for manually selected filter keywords). This seems an indication that the vocabulary and topics of microblogging are different from those found in the Web. Our experiments in Section 4 corroborate this finding.

3.2   Upper Bound Performance of Winner-Takes-All Strategy

In the hypothesis that the tweet stream will be skewed (most tweets will be either related or unrelated to the company), a system that, given a company name, simply guesses which of the two situations is more likely and applies a "winner takes all" classification, may already reach a high average accuracy[3].

---

[3] Of course this is not a solution to the problem from a practical perspective; if there are just a few relevant tweets, we will miss them and we will have no data to analyze. But knowing the majority class may allow us, for instance, to select that class for a tweet in the absence of other relevant signals.

Let us consider an upper bound performance for this strategy, in which this decision is always taken correctly. Results on the WePS-3 testbed are shown in Figure 2b. All points are placed on the upper part of the diagonals, because they are the best possible individual choice between "all positive" or "all negative" for each of the test cases. The average accuracy is 0.80, which is almost comparable with the best manual WePS-3 participation, that obtained an accuracy of 0.83 [7], and superior to the best automatic system of the WePS-3 campaign (with 0.75 accuracy) [8]. These results confirm that an accurate determination of the majority class can be crucial to solve the problem.

Notice that the distribution of related ratios across company names in the WePS-3 corpus streams was enforced to a nearly uniform distribution [1]. In a real scenario, the dots in the fingerprint visualization would be more concentrated in the left and right areas, the effect of predicting the majority class would be even stronger, and therefore the upper bound in real world conditions would be even higher.

3.3   Upper Bound Combination of Strategies

An interesting question is whether filter keywords are enough to predict the majority class. The leftmost graphs in Figure 3 show the results when tagging all tweets for a given company as related or non related depending on the majority class in oracle and manual keywords respectively. Notably, the resulting graph for oracle keywords is almost identical to the *winner-takes-all* upper bound, and results in the same overall accuracy (0.80). Manual keywords, on the other hand, tend to behave as the "all related" baseline, with an average accuracy of 0.61. The reason is probably that there are less manual negative keywords than positive.

The central graphs show the results for the *winner-takes-remainder* strategy, which consists of applying the winner-takes-all strategy only to those tweets that were not previously classified by some of the filter keywords. This straightforward combination of filter keywords and winner-takes-all strategies gives 0.85 accuracy for 20 oracle keywords.

It is also interesting to compare the *winner-takes-all* and the *winner-takes-remainder* approaches with a standard bootstrapping method: we have represented tweets as bag-of-words (produced after tokenization, lowercase and stopword removal) with a binary weighting function for the terms in the instance (occurrence); then we have employed a C4.5 decision tree classification model[4]. For each test case, we use the tweets retrieved by the keywords as seed (training set) in order to classify automatically the rest of tweets. The obtained average accuracy is 0.87 for oracle keywords, only 0.02 absolute points above the *winner-takes-remainder* approach; the graph shows that the improvement resides in cases with a related ratio around 0.5, i.e. the cases where it is more likely to have enough training samples for both classes. Table 1 displays results

---

[4] We also tried with other machine learning methods, such as linear SVM and Naive Bayes, obtaining similar results.

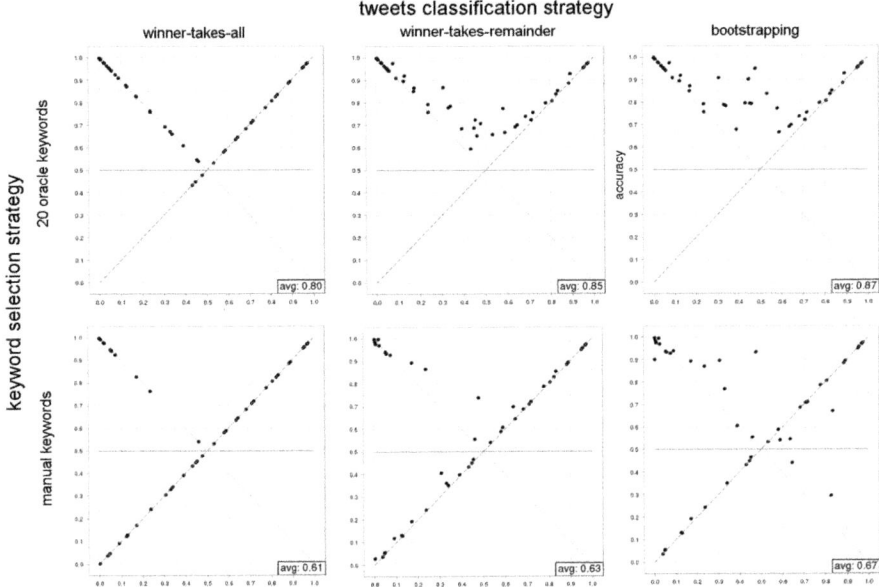

Fig. 3. Fingerprints for each of the three different classification strategies when applying 20 oracle keywords or manual keywords

Table 1. Quality of the three different classification strategies when applying oracle/manual filter keywords

keyword selection strategy	seed set recall	accuracy	overall accuracy winner-takes-all	winner-takes-remainder	bootstrapping
5 oracle keywords	28.45%	1.00	0.80	0.81	0.81
10 oracle keywords	39.97%	1.00	0.80	0.83	0.85
15 oracle keywords	47.11%	1.00	0.80	0.84	0.86
20 oracle keywords	52.55%	1.00	0.80	0.85	0.87
manual keywords	14.61%	0.86	0.61	0.63	0.67

for different amounts of filter keywords: the bootstrapping strategy ranges from 0.81 (with five keywords) up to 0.87 with 20 keywords.

4 Automatic Discovery of Filter Keywords

Our goal is to discover automatically the terms which are most strongly associated to the company name (*positive* filter keywords) or to the alternative meanings of the company name (*negative* keywords), and to discard those which are not discriminative (*skip* terms). To that end, we first discuss a number of potentially helpful term features, and then apply various machine learning and heuristic algorithms that operate on those features.

Table 2. Notation used to describe term features

Item	Description
t, t_i	term
c	(ambiguous) name that identifies a company (e.g. jaguar)
T	set of tweets in the WePS-3 collection[a]
T_c	set of tweets in the collection for a given company name c.
$df_t(T_c)$	document frequency of term t in the collection T_c.
$df_{web}(q)$	number of total hits returned by the Yahoo! Search BOSS API (http://developer.yahoo.com/search/boss/) for the query q.
M	an approximation of the size of the search engine index $(30 \cdot 10^9)$.
$domain_c$	domain of the web site given as reference for the company c.
$wikipedia(q)$	set of Wikipedia pages returned by the MediaWiki API (http://www.mediawiki.org/wiki/API) for the query q.
$dmoz(q)$	set of items (composed by an URL, a title, a description and a category) returned by searching q on the Open Directory Project (http://www.dmoz.org/search)

[a] For each company name, the dataset to which its belongs is only used (either training or test dataset).

4.1 Term Features

Table 2 summarizes the notation used to explain the features. The terms represented are those that are not stopwords and appear at least in five different tweets on T_c, given a company c. For each of these terms, 13 features grouped in three types were computed: *Collection-based features* (*col_**), *Web-based features* (*web_**), and *Features expanded by co-occurrence* (*cooc_**): Given a feature f, a new feature (meant to address the sparseness problem) is computed as the Euclidean norm (1) of the vector composed by $f_{t_i} * w(t, t_i)$ for the five terms most co-occurrent with t in the set of tweets T_c (2), where f_{t_i} is the value of feature f for the term t_i and $w(t, t_i)$ is the the grade of co-occurrence between the terms t and t_i.

$$cooc\_agg(t, f) = \sqrt{\sum_{i \in cooc_t} (f(t_i) * w(t_i))^2} \qquad (1)$$

$$cooc_t = \text{ set of the five most co-occurrent terms with } t. \qquad (2)$$

Table 3 describes the 13 features computed to represent the terms.

4.2 Keyword Discovery

We experiment with three approaches to keyword discovery. The first one, *machine learning*, consists of training a positive-negative-skip classifier over the training corpus in WePS-3 by using the features described previously. We combine two classifiers; positive versus others, and negative versus others. Then, we state one threshold for each classifier. Terms which are simultaneously under/over both thresholds are tagged as skip terms. We have tried with several machine learning methods using Rapidminer [5]: Multilayer Perceptron with Backpropagation (Neural Net), C4.5 and CART Decision Trees, Linear Support

Table 3. Features computed to represent the terms

Feature	Description
$col\_c\_df = df_t(T_c)/|T_c|)$	Normalized document frequency over the collection of the company tweets.
$col\_c\_specificity = df_t(T_c)/df_t(T)$	Ratio of document frequency on the company collection T_c over the document frequency on the collection T.
$col\_hashtag$	Number of occurrences that the term appears as hashtag on T_c.
$web\_c\_assoc = \dfrac{df_{web}(t\ OR\ c)/df_{web}(c)}{df_{web}(t)/M}$	Represents the association, according to the search counts, between the term t and a company name c.
$web\_c\_ngd = \dfrac{\max(\log(f(c)),\log(f(t))-\log(f(t\ AND\ c))}{M-\min(\log(f(t)),\log(f(c))}$ where $f(x) = df_{web}(x)$	The Normalized Google Distance [2] (applied to the Yahoo! search engine) between a term t and a company name c
$web\_dom\_df = \dfrac{df_{web}(t\ AND\ site:domain_c)}{df_{web}(site:domain_c)}$	Normalized document frequency of the term in the website of the company.
$web\_dom\_assoc = \dfrac{web\_dom\_df}{df_{web}(t)/M}$	Analogous to web_c_assoc, using the website domain instead of the company name c.
web_odp_occ	Number of occurrences of the term on all the items retrieved in $dmoz(domain_c)$.
web_wiki_occ	Number of occurrences of the term on the first 100 results in $wikipedia(domain_c)$.
$cooc\_c\_assoc$, $cooc\_c\_ngd$, $cooc\_dom\_df$, $cooc\_dom\_assoc$	Features col_c_assoc, web_c_ngd, web_dom_df and web_dom_assoc expanded by co-occurrence.

Vector Machines (SVM) and Naive Bayes. According to the ROC curves of those classifiers, Neural Nets are slightly superior.

The second approach (*heuristic*) is inspired by an analysis of the signal provided by each of the features, and consists of a heuristic which involves the two most informative features. First, we define a threshold to remove skip terms which are not specific enough with respect to the tweet set (using the *col_c_spec* feature). Then we state, using the feature that measures association with the website (*cooc_dom_assoc*) a minimal value to consider a keyword positive and a maximal value under which keywords are considered negative. These three thresholds were manually optimized using the training data set. Finally, the *hybrid* approach consists of applying machine learning as in the first method, but using only the two features applied in the heuristic approach.

After detecting the filter keywords automatically, we classify the tweets that contain only negative or only positive keywords. Then, in order to classify the rest of tweets, we apply the same three alternative methods described in Section 3: winner-takes-all, winner-takes-remainder and bootstrapping.

Table 4 shows the results of our experiments. The best automatic system (machine learning to discover keywords and bootstrapping with the tweets annotated using that keywords) gives an accuracy of 0.73, which is much better than using manual keywords (0.67) and close to the best automatic result in the competition (0.75). Note that the winner-takes-remainder strategy, which is a direct combination of the two strategies (filter keywords plus winner-takes-all)

Table 4. Results for automatic keyword detection strategies

keyword selection strategy	seed set recall	seed set accuracy	overall accuracy winner-takes-all	winner-takes-remainder	bootstrapping
m. learning	57.81%	0.72	0.64	0.68	**0.73**
heuristic	27.42%	0.79	0.64	0.65	0.71
hybrid	38.64%	0.78	0.70	0.72	0.72

reaches 0.72; this is a strong baseline which shows that both signals are indeed useful to solve the problem.

Discovery of filter keywords has proved to be challenging using signals from the Web: the accuracy of the resulting seed set ranges between 0.72 and 0.79, with a large recall only in the case of the machine learning strategy. As with the manual selection of keywords, the biggest problem is finding suitable negative keywords. Overall, this result reinforces the conclusion that the characterization of companies in Twitter, in terms of vocabulary, is different from the characterization in Web pages.

5   Discussion and Conclusions

This paper makes a few novel contributions to the problem of company name disambiguation in microblog entries: (i) we introduce a *fingerprint* visualization technique to understand and compare systems' behavior (ii) we show that finding *filter keywords* and determining the majority class in a set of tweets are two relevant sources of information to solve the problem; and (iii) we have seen that the vocabulary that characterizes a company in Twitter is substantially different from the vocabulary associated to the company in its home page, in Wikipedia, and in the Web at large.

We have seen that in the WePS-3 dataset, five optimal keywords directly classify around 30% of the tweets in average (providing valuable information to characterize the positive and negative classes), and assigning the true majority class to all tweets (*winner-takes-all* strategy) gives .80 accuracy (which is higher than the best published automatic result on this test collection).

We have also set upper and lower bound classification performances using only these two signals to disambiguate tweets:

- **Upper bound:** Using 20 *oracle* (perfect) filter keywords and a winner-takes-all guess based on the majority class in the tweets that contain the filter keywords we reach .85 accuracy without any additional information or algorithmic machinery.
- **Lower bound:** If we replace oracle keywords with automatically extracted filter keywords we reach .72 accuracy, which is close to the best published result on WePS-3 dataset (.75).

Remarkably, the contribution of the naive winner-takes-all strategy to the lower bound is not easy to improve. If we replace it with a Machine Learning process (using tweets with filter keywords as training set) we reach .73 accuracy, which represents only a 1.4% improvement. Our experiments also indicate that, although both signals are useful to solve the problem, extracting filter keywords and guessing the majority class are challenging tasks. In the case of filter keywords, we have seen that the vocabulary that characterizes a company in the Web has only a moderate overlap with its microblogging counterpart. In fact, a manual keyword selection from Web sources only reaches 14.61% coverage with 0.86 accuracy. Descriptions of the company in its home page, in Wikipedia, and in the Web at large, provide useful but weak signals to the problem of filter keyword identification, and it is far from trivial to build an accurate classifier with them. Negative keywords (those which signal tweets which are not related to the company) are particularly hard to discover automatically. Guessing the majority class seems to be, apparently, a simpler problem, but in practice it turns out to be equally hard.

Our future work naturally addresses two questions: how to find filter keywords (and determine the majority class) more accurately, and how to integrate both types of information in a full-fledged algorithmic solution to the problem. Given that we have only used these two signals to disambiguate so far, we believe that merging them with other useful signals should provide an optimal solution to the problem.

References

1. Amigó, E., Artiles, J., Gonzalo, J., Spina, D., Liu, B., Corujo, A.: WePS-3 Evaluation Campaign: Overview of the Online Reputation Management Task. In: CLEF 2010 Labs and Workshops Notebook Papers (2010)
2. Cilibrasi, R.L., Vitanyi, P.M.: The google similarity distance. IEEE Transactions on Knowledge and Data Engineering (2007)
3. García-Cumbreras, M.A., García-Vega, M., Martínez-Santiago, F., Peréa-Ortega, J.M.: SINAI at WePS-3: Online Reputation Management. In: CLEF 2010 Labs and Workshops Notebook Papers (2010)
4. Kalmar, P.: Bootstrapping Websites for Classification of Organization Names on Twitter. In: CLEF 2010 Labs and Workshops Notebook Papers (2010)
5. Mierswa, I., Wurst, M., Klinkenberg, R., Scholz, M., Euler, T.: YALE: Rapid prototyping for complex data mining tasks. In: SIGKDD 2006: Proceedings of the 12th International Conference on Knowledge Discovery and Data Mining (2006)
6. Tsagkias, M., Balog, K.: The University of Amsterdam at WePS3. In: CLEF 2010 Labs and Workshops Notebook Papers (2010)
7. Yerva, S.R., Miklós, Z., Aberer, K.: It was easy when apples and blackberries were only fruits. In: CLEF 2010 Labs and Workshops Notebook Papers (2010)
8. Yoshida, M., Matsushima, S., Ono, S., Sato, I., Nakagawa, H.: ITC-UT: Tweet Categorization by Query Categorization for On-line Reputation Management. In: CLEF 2010 Labs and Workshops Notebook Papers (2010)

Automatic Annotation of Bibliographical References for Descriptive Language Materials

Harald Hammarström*

Max Planck Institute for Evolutionary Anthropology
Department of Linguistics
Deutscher Platz 6
D-04 150 Leipzig
Germany
h.hammarstrom@let.ru.nl

Abstract. The present paper considers the problem of annotating bibliographical references with labels/classes, given training data of references already annotated with labels. The problem is an instance of document categorization where the documents are short and written in a wide variety of languages. The skewed distributions of title words and labels calls for special carefulness when choosing a Machine Learning approach. The present paper describes how to induce Disjunctive Normal Form formulae (DNFs), which have several advantages over Decision Trees. The approach is evaluated on a large real-world collection of bibliographical references.

Keywords: Document Categorization, Supervised Learning, Cross-Lingual Information Retrieval, Decision Trees, Language Documentation.

1 Introduction

LangDoc is a large-scale project to list bibliographical references to descriptive materials to all of the ca 7 000 languages of the world [1]. The present collection contains nearly 160 000 such references.

A linguist, typically a typologist searching/browsing through references, would want the collection *systematically* annotated with metadata, such as the identity of the [target-]language(s) the reference treats, the geographical location country/continent, the content-type of the document the reference refers to (e.g., (full-length) grammar, grammar sketch, dictionary, phonological description) and so on.

The present collection of 160 000 references comes from a variety of sources, some of which are already annotated with metadata, and this can be exploited in terms of supervised learning.

For example, a bibliographical reference to a descriptive work may look as follows:

---

* The author wishes to thank Sebastian Nordhoff, Martin Haspelmath, Guillaume Ségérer, Jouni Filip Maho and Alain Fabre for various kinds of input relevant to the present study.

P. Forner et al. (Eds.): CLEF 2011, LNCS 6941, pp. 62–73, 2011.

Schneider, Joseph. 1962. *Grammatik der Sulka-Sprache (Neubritannien)* (Micro-Bibloteca Anthropos 36). Posieux: Anthropos Institut.

This reference happens to describe a Papuan language called Sulka [sua], it is a grammar (rather than a dictionary, grammar sketch etc.), and is further tagged with Oceania (macro-area) and Papua New Guinea (country). This example reference is written in German (i.e., the [meta-]language that the publication, and therefore reference, is written in – not the [target-]language that the publication aims to describe).

Now suppose we are given a new bibliographical reference which has no annotation. We would like to automatically annotate it with identity, type and whatever other labels are justified, given the training data consisting of already annotated references. For example, many titles in the training data will contain the word "Grammatik" and be annotated with grammar, those few which have the word "Neubritannien" will likely be annotated with Oceania and Papua New Guinea and so on.

Unfortunately, the problem is not as simple as checking for statistically significant keywords.

2   Problem Statement

The problem at hand can be seen as a special case of a more general Information Extraction problem with the following characteristics.

- There is a set of natural language objects O
- There is a fixed set of categories C
- Each object in O belongs to zero or more categories, i.e., there is a function $Z : O \rightarrow Powerset(C)$
- The task is to find classification function f that mimics Z.

The special case we are considering here is such that:

- Each object in O contains a small amount of text, on the order of 100 words
- The language of objects in O varies across objects, i.e., not all objects are written in the same language
- $|C|$ is large, i.e., there are many categories (in our case 5 471 + 14 + 6 = 5 491 classes, see Table 1)
- $|Z(o)|$ is small for most objects $o \in O$, i.e., most objects belong to very few categories
- Most objects $o \in O$ contain a few tokens that near-uniquely identifies $Z(o)$, i.e., there are some words that are very informative as to category, while the majority of tokens are very little informative. (This characteristic excludes the logical possibility that each token is fairly informative, and that the tokens *together*, on an equal footing, serve to pinpoint category.)

3   The Present Dataset

As mentioned already, the present collection contains nearly 160 000 references. The (meta-)languages of the references are (in descending order of frequency): English, German, French, Spanish, Portuguese, Russian, Dutch, Italian, Chinese, Indonesian, Japanese, Nepali, Afrikaans, Thai, Hindi, Turkish, Arabic, Georgian, Urdu, Bulgarian, Swedish, Finnish, Danish, Norwegian, Assamese, Swahili, Burmese, Polish and a few other languages which are represented in less than 10 references. Table 1 shows the incidence of various types of annotation already present. The annotation stems from the various sources of the collection (see [1] for more information on the composition and provenance of the catalogue), and some inconsistencies can therefore be expected. There is also some further largely idiosyncratic annotation from subparts of the collection (e.g., country, keywords, shelf-mark, ...) which is excluded from the present study since it overlaps in function with those selected for Table 1 (but could in principle be addressed with the same methods as in the present study).

Table 1. Size of the present database of references and incidence of annotation already in place

Annotation type	# different labels	# annotated references
Macro-area	6	121 296
Content-type	14	15 236
Target-language	5 471	88 978
Total # of annotated references		158 498

There are six possible macro-area labels, but they are not mutually exclusive. For example, a reference to a publication dealing with Africa as well as South America, should be labelled with both. Similarly, there are 5 471 different labels for target-language, and it is logically possible for a publication to refer to any subset of these, though, in practice, most references tend to target only one or a few languages. Type refers to the type of descriptive data, such as grammar, dictionary, grammar sketch, wordlist, ethnographic work, texts and eight others. The typology of type is, for historical reasons, somewhat ad hoc, but nevertheless useful to the target community of searchers. Some types logically exclude each other, e.g., a reference cannot both be a grammar and a grammar sketch, while others are compatible, e.g., a single book can contain both a grammar and a dictionary.

As to the size of the task at hand, Table 1 shows that, for example, out of the 158 498 references, 121 296 of them are already annotated (by a human) as to macro-area, but the remainder, 37 202 are in need of macro-area labels. (It is assumed that references which are already annotated with a certain kind of annotation are not in need of *more* annotation of the same kind.)

At first, this problem, i.e., reference annotation by keyword triggers, might seem like a very easy problem – just find title words which are statistically overrepresented with an annotation label in the training data, and then label

Table 2. Two example labels with some potential trigger words and their ability to "select" the respective labels

# references contains		grammar # with `grammar`	label precision recall	
162	"grammatik"	91	0.56	0.068
668	"der"	137	0.21	0.103
84	"grammatik", "der"	48	0.57	0.036
1	"sulka"	1	1.00	0.001
# references contains		Sulka [sua] # with `Sulka [sua]`	label precision recall	
1	"sulka"	1	1.00	0.16
668	"der"	4	0.01	0.67

new instances as such words occur in their titles. However, there are a few reasons why it is not that simple.

- A label may be signalled by more than one word, e.g., "kurzgefaßte grammatik" signals `grammar sketch` rather than `grammar` (not both!).
- It is not given which keyword(s) signal which label(s), e.g., from the example above, is it "Grammatik", "der" or "Grammatik der" (all of them statistically significant[1]) that signals `grammar`?
- Some labels are very common (and thus have frequent trigger words) while other labels are very uncommon (and thus their trigger words are very uncommon). This means that simple frequency thresholds cannot be used to rule out useful trigger words.
- Typically, a small set of trigger words "account" for an annotation label, i.e., no single one of them has a high recall with its label, but together they do.

For example, among 15 236 references annotated for content-type 19 921 distinct word types are present. 3 220 have the label `grammar` and 6 have the label `Sulka [sua]`. Table 2 shows that even a words like "grammatik" and "sulka" have rather low recall for their respective "true" labels, and if we combine precision and recall, there is some serious competition of (combinations with) spurious trigger words such as "der".

We will explore some Machine Learning ideas to come up with a solution tailored to the particularities of this classification problem.

4   Related Work

The approach in the present paper generalizes the method of [2] to annotate bibliographical references with only uncommon labels. We are not aware of any other work specifically targeting the annotation of bibliographical references based on the text of the reference itself.

---

[1] There is a reason why a word such as "der" is overrepresented with a label such as `grammar`. The label grammar is triggered by words like "Grammatik", but because of the rules of the (in this case) German language, "der" is also caused to be in title of (most) references with "Grammatik" in them.

The problem, however, has a clear analogue in Information Retrieval in the following sense. Typically, the task is to find a set of relevant documents given a document collection and a query. On the other hand, if we equate documents with the text of a bibliographical reference, and the set of relevant documents and the set of references with a certain label, then the problem addressed in this paper is to find the query given the document collection and the set of relevant documents. As such the problem has been addressed in terms of word-space models [3], and special focus has been on the special case of sentiment analysis [4]. Such work also includes principled approaches to the multi-lingual situation [5,6], though often relying on existing lexical resources, e.g., dictionaries. Such approaches scale well to large collections, but are otherwise imperfectly suited to the specifics of the problem in the present paper. First, we do not have access to dictionaries for the full range of languages featured in the present collection. Second, the techniques described output large word-probability tables which combine evidence from many words in a long document, whereas in the present collection, every document is very short. Third, most techniques described involve human-tuned seed data or thresholds which make the approaches less attractive to work with.

A number of techniques which have been successful for Text Classification (cf. [7], though somewhat dated, the principles outlined therein remain valid) are less well-suited for the present problem. There are dependencies between words that go against the Naive Bayesian assumption, e.g., "grammar" signals the label **grammar** if and only if not occurring with "short", "sketch" etc. Naive Bayes, along with a number of other statistical approaches, have no way of distinguishing which keyword(s) in a title signals signal which label(s) and end up distributing the evidence over all words in the title (which is not fatal, but unnecessary). Extra work is also required in smoothing techniques for infrequent labels. Other statistical techniques work best after text processing such as stopword removal and/or tf-idf-weighting. In the present context, we do not have access to stopword lists for the full range of languages targeted, and there is also the suspicion that what are stopwords in regular prose may not correspond to stopwords in publication titles (the same suspicion can be raised for other enhancements that tap into linguistic structures [8]). Similarly, in the very multilingual setting, tf-idf weighting is significantly crippled, as what are frequent words within *one* language will have only a fraction of their frequency when diluted in large pool of languages (and there may be distorting interferences across languages).

The traditional principled approach to classification of objects with a set of discrete valued features are decision trees [9], which, in addition, are able to "explain" their predictions. ID3 Decision Trees are well-suited for the present problem except that they may become unnecessarily large and that setting a depth-threshold is required. The reason ID3 Decision Trees may become "unnecessarily large" in the present setting is that they are designed to be built complete, e.g., if the attribute "grammaire" is chosen as a branch, the corresponding negated branch must also be present, and both branches must be filled in the next round of iterations. In the present problem setting, we envisage the optimal tree to look more like a rake than a tree. Although general-purpose pruning heuristics to decision trees are widely used (as per C4.5 [10]), a solution specially designed to allow rake-like classifiers obviates the need for thresholds and pruning heuristics.

There are thus good theoretical arguments for re-assuming from the 1990s the approach of rule-induction classifiers [11,12,13] for the particular problem setting addressed here.

5   A DNF Approach

As outlined above, our domain knowledge suggests that a label can be inferred if and only if a suitable combination of words is present/absent in a given publication title. More formally:

- A trigger-signature $t = w_1 \wedge \ldots \wedge w_k \wedge \ldots \neg w_{k+1} \wedge \ldots \wedge \neg w_{k'}$ for a label l is a conjunct formula of negated/un-negated terms, such that if a title contains all the un-negated terms but none of the negated terms, then the label l should be inferred.
- Each label l can have one or more trigger-signatures $t_1, \ldots, t_n$

For example, one trigger for the label grammar might be $\{grammar, \neg sketch\}$, and the full set of triggers for grammar might contain $\{grammar, \neg sketch\}$, $\{grammaire\}$, $\{complete, description\}$, $\{phonologie, morphologie, syntax\}$ and so on. Since titles are short (less than 20 words or so), we envisage triggers to be short.

In other words, a classifier (one for each label) can be described as a boolean formula in DNF, where each disjunct corresponds to a trigger. Moreover, each disjunct can be expected to be relatively short.

Thus, all we need to do is to search for a formula in DNF form which can be expected to have only short disjuncts and which is preferably short (in its number of disjuncts). Thus, a simple algorithm is to start from an empty formula and build it larger as accuracy increases with respect to a label in the training data. One can build a formula larger either:

i   by adding a negated/un-negated term to one of its disjuncts (replacing that disjunct[2]), or
ii by adding a new disjunct, inhabited by a negated/un-negated literal.

Since we are interested in both high precision and high recall, a natural way to measure accuracy is f-score.

The following notation will be used:

- $d_i \subseteq \Sigma^*$ be a document, i.e., a set of strings
- $D = \{d_1, \ldots, d_n\}$ be a set of documents

---

[2] To keep an updated and un-updated disjunct is superflous since $A = A \vee (A \wedge B)$.

- $W_D = \bigcup d_i$ be the set of terms of a set of documents
- $L_D(l) = \{i | d_i \text{ has label } l\}$ be the subset of documents with label l
- $c = \bigvee t_j$ be a DNF boolean formula
- $c_D = \{i | c \text{ is true for } d_i\}$ be the subset of documents whose terms satisfy a boolean formula c
- $Precision_D(c, l) = |c_D \cap L_D(l)| / |c_D|$
- $Recall_D(c, l) = |c_D \cap L_D(l)| / |L_D(l)|$

The training algorithm can be described as follows:

1. Start with a label l, a document collection D and an empty formula c
2. Form sets of candidate formulae

$$C' = \{c \vee w | w \in W_D\} \cup \{c \vee \neg w | w \in W_D\}$$
$$C'' = \{ins(w, t_j, c) \vee t_j | w \in W_D, t_j \text{ of } c\} \cup$$
$$\{ins(\neg w, t_j, c) \vee t_j | w \in W_D, t_j \text{ of } c\}$$

where $ins(x, t_j, c)$ means "replace t_j with $t_j \wedge x$ in the formula c", e.g., $ins(c, t_2, (a \wedge \neg b) \vee (a)) = (a \wedge \neg b) \vee (a \wedge c)$.
3. Compute $c' = argmax_{c' \in C' \cup C''} \text{f-score}_D(c', l)$
4. If c' equals c finish, otherwise set c to c' and iterate from step 2

6   Experiments

6.1   Experiment Design

Since the labels are largely independent, we trained one DNF for each label. To classify an unseen reference, we test it with all DNFs in parallel, and label it accordingly.

As noted already, the labels fall into three classes: Target language, content-type and macro-area. For each class, we randomly selected 1000 previously annotated references to use as a test set. These were set apart from the beginning and were never accessed during development.

With the intended search audience in mind, we believe that precision is more important than recall, especially since there are catch-all labels based on geography that make up for some loss of recall. Consequently, all experiments were run to optimize the $F_{0.5}$-score of a DNF, where precision is twice as important as recall [14].

All titles are in roman script or have transcriptions. All title words were lowercased and all diacritics and accents were removed.

6.2   Results

Overall results, in numbers, grouped by label class, are shown in Table 3. The overall f-score, precision and recall figures are based on numbers of labels (rather than numbers of references, since many references have more than one label of the same class).

Table 3. Overall accuracy of the DNF approach, grouped by label class

Label Class	# labels	# training refs	$|W_D|$	Overall $F_{0.5}$	Overall Precision	Overall Recall
Macro-area	6	121 296	117 213	0.60	0.57	0.76
Content-type	14	15 236	19 921	0.57	0.59	0.51
Target-language	5 471	88 978	83 828	0.80	0.85	0.66
	5 491			0.70	0.70	0.69

Table 4. Accuracy for macro-area labels

Label	# in training data	$F_{0.5}$	Precision	Recall
Australia	6 988	0.76	0.91	0.46
Eurasia	10 579	0.50	0.61	0.28
North America	2 311	0.45	0.50	0.31
Africa	69 734	0.57	0.52	0.95
Oceania [except Australia]	3 681	0.46	0.51	0.33
South America	29 100	0.62	0.61	0.66
	122 393	0.60	0.57	0.76

It is instructive to look closer at the results for macro-area labels in particular, shown in Table 4

The DNF for Australia was extracted straightforwardly as

$$aboriginal \lor australian \lor australia \lor warlpiri \lor queensland \lor aborigines \lor$$
$$arnhem \lor pitjantjatjara \lor torres \lor nyungan \lor (wales \land new \land south) \lor$$
$$arrernte \lor yolngu \lor dyirbal \lor kriol \lor kimberley \lor york$$

i.e., some geographical names and some names of languages/families prominently present in Australia. The DNF for Eurasia is similar

$$thai \lor tibetan \lor jazyka \lor burmese \lor viet \lor tai \lor yu \lor vietnamese \lor khmer \lor$$
$$slovar \lor chinese \lor iazyke \lor siamese \lor nepal \lor hmong \lor miao \lor hindi \lor tibeto \lor$$
$$phasa \lor india \lor tieng \lor slov \lor japanese \lor thailand \lor grammatika \lor burman$$

except that here we also have a 'grammatika', 'slovar' and 'iazyke', i.e., which are Russian words, indirectly indicating Eurasia only since the vast majority of most Russian works target Eurasian languages. The DNF for North America has a list of specific language/family names but only one country name "Mexico". The DNF for Africa is different

$$(\neg america \land \neg o \land \neg american \land \neg story \land \neg lengua \land \neg australia \land \neg indians \land$$
$$\neg do \land \neg australian \land \neg new \land \neg thai \land \neg grammar \land \neg i \land \neg review \land \neg el \land$$
$$\neg aboriginal \land \neg e \land \neg los \land \neg del \land \neg y) \lor africa \lor swahili \lor bantu \lor hausa \lor$$
$$congo \ldots$$

containing a large trigger of negated literals. This is presumably because the label Africa is numerically dominant. This trigger also accounts for the unusually high recall figure. The DNF for South America contains a large number of common

Table 5. Accuracy for content-type labels

Label	# in training data	$F_{0.5}$	Precision	Recall
handbook/overview	4 549	0.60	0.61	0.58
grammar	3 216	0.63	0.65	0.55
comparative-historical	2 992	0.50	0.49	0.59
grammar sketch	2 519	0.38	0.42	0.26
ethnographic work	1 886	0.59	0.59	0.60
wordlist	1 807	0.48	0.57	0.29
dictionary	926	0.78	0.83	0.64
study of a specific feature	626	0.43	0.46	0.34
bibliographic	550	0.72	0.76	0.61
very small amount of information	541	0.67	0.68	0.63
sociolinguistic	493	0.54	0.55	0.51
phonology	347	0.68	0.75	0.51
text	149	0.94	0.90	0.90
dialectology	124	0.81	0.80	0.88
	20 725	0.57	0.59	0.51

Spanish and Portuguese words, a reflection of the fact that words with Spanish and Portuguese are concentrated to South America.

Results for content-type labels, shown in Table 5 are similar, despite the much smaller size of the training set. All DNFs extracted are short (less than 50 disjuncts) and contain little of surprise. For example, the DNF for grammar is

$grammar \lor grammaire \lor jazyk \lor gramatica \lor grammatik \lor grammatika \lor$
$description \lor course \lor parlons \lor syntax \lor manuel \lor jazyka \lor spraakkunst \lor$
$(phonologie \land morphologie) \lor grammatica \lor descriptive \lor manual \lor arte \lor$
$dialekt \lor handbook$

and the DNF for phonology contains the trigger signature $\ldots \lor (phonologie \land \neg morphologie) \lor \ldots$, precisely as expected.

Inspection shows that a lot of errors come from the fact that the labels grammar and grammar sketch are not quite distinguishable by title words alone.

For the labels in the target-language class, their frequency in the training data ranges from 1 440 of Hausa [hau] to 1 025 labels, e.g., Wurrugu [wur], with frequency 1. In the test set of 1000 references, there were an additional 211 target-language label types [221 label tokens] which do not appear even once in the training data. It is impossible to find such labels given the training data, so we also present recall figures which are adjusted upwards. Most labels in the target-language class do not appear in the 1000 item test set, wherefore we also show the precision and recall figures on the training data (as some kind of indication of the power of DNFs for these labels). Table 6 shows these results on the basis of all label tokens together – the labels are far too many to inspect on an individual basis.

Target-language labels are easy to capture with DNFs, because of the typical title contains (at least one) near-unique identifier. Top, median and bottom frequency examples are shown in Table 7.

Table 6. Overall accuracy for target-language labels

Data Set	# label tokens	$F_{0.5}$	Precision	Recall
Test Data	1 292	0.80	0.85	0.66
Test Data Adjusted	1 071	0.83	0.85	0.79
Training data	118 065	0.87	0.93	0.73

Table 7. Example DNFs for target-languages labels on the training data set

Label	# label tokens	$F_{0.5}$	Precision	Recall	DNF
Hausa [hau]	1 440	0.94	0.99	0.80	$hausa \lor haoussa \lor$ $haussa \lor hausaland \lor$ $hawsa \lor xausa$
Nisenan [nsz]	6	0.96	1.00	0.83	$nisenan$
Wurrugu [wur]	1	1.00	1.00	1.00	$wurrugu$

Errors in precision and recall come mainly from cases where one publication treats several languages, but the title does not list them, e.g., *Languages of the Eastern Caprivi*.

All DNFs for target-language labels are short (less than 15 disjuncts). In a fair amount of cases, a trigger signature contains a seemingly superfluous word, e.g., *the* ∧ *mayi*, transparently because adding one word to *mayi* makes the title unique in the training set. Presumably, there are several words which suffice equally well, but in reality, a specific one is appropriate. A future tweak could target this pattern. Similarly, there are cases of hapax words, e.g., "syllabics", which are not language names, yet since show up with only one target-language label, they are indistinguishable from a true language name in the present approach. Since such words are rare, little or no classification errors can be expected to result from such spurious language identifiers.

6.3   Discussion

The accuracy in the results obtained will certainly be useful in that it will save a lot of human annotation time. But since automatic annotation is imperfect and incomplete, a human will still need to browse the results and correct errors.

The performance is significantly better than ID3 Decision Trees [9] whose performance on this problem (with one tree per label, as with DNFs) yields much larger trees for the same f-scores, and require threshold (tree-height) settings for training to stop. For example, ID3, on the same training data, produces the following tree for the label `grammar`:

which has only $F_{0.5} \approx 0.42$ (precision 0.55, recall 0.22) on the same test set. The similar-sized DNF (shown above) has $F_{0.5} \approx 0.63$ (precision 0.65, recall 0.55). Cutting the tree deeper only has marginally higher scores. This is likely due to the fact that the tree is designed to be dense, and so has to create a number of inefficient branches, whereas the DNF can mimic a sparse tree, which better fits the problem. A Naive Bayesian classifier for the label grammar achieves only $F_{0.5} \approx 0.35$ (precision 0.34, recall 0.45) on the same test set.

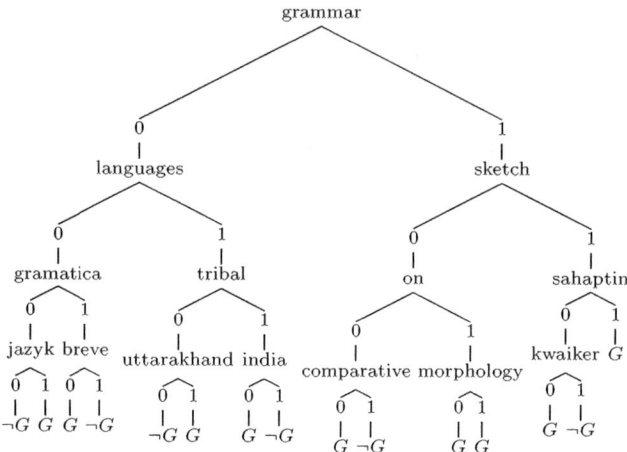

The algorithm for finding DNFs is subject to falling into local minima – indeed it is the only obstacle to overfitting. However, Since the output DNFs correspond well to intuitions, we have not investigated to what extent there are globally more accurate DNFs than the ones found by the algorithm.

Training DNFs is worst-case quadratic in $|W_D|$. Given the search space with a large W_D in the present case, this is rather slow. It is likely that intelligent filtering of W_D may significantly reduce it, but since training speed is not an issue, this has not been explored.

A drawback of the present approach is that non-boolean attributes cannot be elegantly integrated. For example, it seems likely that the number of pages (of the work that a reference points to) is highly relevant for the difficult decision between the labels grammar and grammar sketch. In our current reference collection, page numbers are not systematically present, so we are unable to check this matter thoroughly anyway.

The output formulae are readily interpretable to a human, thus the classifier annotating a new reference can "explain" its result.

7   Conclusion

We have presented a principled approach to supervised document categorization on very short documents written in a variety of languages. The present approach has advantages in elegance over alternative machine learning methods and can cope equally with common and uncommon categories, i.e., with sparse amounts of training data. The approach is thoroughly evaluated on a collection of bibliographic references and will be used in practice.

References

1. Hammarström, H., Nordhoff, S.: Langdoc: Bibliographic infrastructure for linguistic typology. Oslo Studies in Language, 14 (in press, 2011)
2. Hammarström, H.: Automatic annotation of bibliographical references with target language. In: Proceedings of MMIES-2: Workshop on Multi-source, Multilingual Information Extraction and Summarization, ACL, pp. 57–64 (2008)
3. Sahlgren, M.: The Word-Space Model: Using distributional analysis to represent syntagmatic and paradigmatic relations between words in high-dimensional vector spaces. PhD thesis, Stockholm University, Stockholm (2006)
4. Huang, X., Croft, W.B.: A unified relevance model for opinion retrieval. In: CIKM 2009: Proceeding of the 18th ACM Conference on Information and Knowledge Management, pp. 947–956. ACM, New York (2009)
5. Lavrenko, V., Choquette, M., Croft, W.B.: Cross-lingual relevance models. In: SIGIR 2002: Proceedings of the 25th Annual International ACM SIGIR Conference on Research and Development in Information Retrieval, pp. 175–182. ACM, New York (2002)
6. Zhang, D., Mei, Q., Zhai, C.: Cross-lingual latent topic extraction. In: ACL 2010: Proceedings of the 48th Annual Meeting of the Association for Computational Linguistics, pp. 1128–1137. Association for Computational Linguistics, Morristown (2010)
7. Sebastiani, F.: Machine learning in automated text categorization. ACM Computing Surveys 34, 1–47 (2002)
8. Al Zamil, M.G.H., Can, A.B.: Rolex-sp: Rules of lexical syntactic patterns for free text categorization. Knowledge-Based Systems 24(1), 58–65 (2011)
9. Quinlan, J.R.: Induction of decision trees. Machine Learning 1(1), 81–106 (1986)
10. Quinlan, J.R.: C4.5: programs for machine learning. Morgan Kaufmann, San Francisco (1993)
11. Cohen, W.W.: Fast effective rule induction. In: Proceedings of the Twelfth International Conference on Machine Learning, pp. 115–123. Morgan Kaufmann, San Francisco (1995)
12. Clark, P., Niblett, T.: The cn2 induction algorithm. Machine Learning 3, 261–283 (1989)
13. Sever, H., Gorur, A., Tolun, M.R.: Text Categorization with ILA. In: Yazıcı, A., Şener, C. (eds.) ISCIS 2003. LNCS, vol. 2869, pp. 300–307. Springer, Heidelberg (2003)
14. van Rijsbergen, C.J.: Information Retrieval, 2nd edn. Butterworths, London (1979)

A Language-Independent Approach to Identify the Named Entities in Under-Resourced Languages and Clustering Multilingual Documents

N. Kiran Kumar, G.S.K. Santosh, and Vasudeva Varma

International Institute of Information Technology, Hyderabad, India
{kirankumar.n,santosh.gsk}@research.iiit.ac.in, vv@iiit.ac.in

Abstract. This paper presents a language-independent Multilingual Document Clustering (MDC) approach on comparable corpora. Named entites (NEs) such as persons, locations, organizations play a major role in measuring the document similarity. We propose a method to identify these NEs present in under-resourced Indian languages (Hindi and Marathi) using the NEs present in English, which is a high resourced language. The identified NEs are then utilized for the formation of multilingual document clusters using the Bisecting k-means clustering algorithm. We didn't make use of any non-English linguistic tools or resources such as WordNet, Part-Of-Speech tagger, bilingual dictionaries, etc., which makes the proposed approach completely language-independent. Experiments are conducted on a standard dataset provided by FIRE[1] for their 2010 Ad-hoc Cross-Lingual document retrieval task on Indian languages. We have considered English, Hindi and Marathi news datasets for our experiments. The system is evaluated using F-score, Purity and Normalized Mutual Information measures and the results obtained are encouraging.

Keywords: Multilingual Document Clustering, Named entities, Language-independent.

1 Introduction

Multilingual Document Clustering (MDC) is the grouping of text documents, written in different languages, into various clusters, so that the documents that are semantically more related will be in the same cluster. The ever growing content on the web, which is present in different languages has created a need to develop applications to manage huge amount of this varied information. MDC has been shown to be a very useful application which plays a major role in managing such varied information. It has got applications in various streams such as Cross-lingual Information Retrieval (CLIR) [1], where search engine takes query in one language and retrieves results in different languages. Instead of providing results as a single long list, the search engine should display them

---

[1] Forum for Information Retrieval Evaluation - http://www.isical.ac.in/~clia/

P. Forner et al. (Eds.): CLEF 2011, LNCS 6941, pp. 74–82, 2011.
© Springer-Verlag Berlin Heidelberg 2011

as list of clusters, where each cluster contains multilingual documents that are similar in content. It encourages users for cluster based browsing which is very convenient for processing the results.

Named entities (NEs) are the phrases that contain names of persons, organizations, locations, times, and quantities. Extracting and translating such NEs benefits many natural language processing problems such as Cross-lingual Information Retrieval, Cross-lingual Question Answering, Machine Translation and Multilingual Document Clustering. Various tools such as Stanford Part-of-Speech (POS) tagger, WordNet, etc., are available to identify the NEs present in high resourced languages such as English. However, under-resourced languages don't enjoy such facility due to the lack of sufficient tools and resources. To overcome this problem we proposed an approach to identify the NEs present in under-resourced languages (Hindi and Marathi) without using any language dependent tools or resources. The identified NEs are then utilized in the later phase for the formation of multilingual clusters. Fig. 1 gives an overview of the proposed approach.

Bilingual dictionaries, in general, don't cover many NEs. Hence, we used a Wiki dictionary [2] instead of bilingual dictionaries to translate Hindi and Marathi documents into English. The Wiki dictionary covers broader set of NEs and is built availing multilingual Wikipedia titles which are aligned using cross-lingual links. During the translation, in order to match a word in a document with dictionary entries, we require lemmatizers to stem the words to their base forms. But as stated earlier, the support of lemmatizers is limited in under-resourced languages. As an alternative, we used Modified Levenshtein Edit Distance (MLED) [3] metric. It solves the problem of 'relaxed match' between two strings, occuring in their inflected forms. This MLED is used in matching a word in its inflected form with its base form or other inflected forms. The rules are very intuitive and are based on three aspects:

1. Minimum length of the two words.
2. Actual Levenshtein distance between the words.
3. Length of subset string match, starting from first letter.

The rest of the paper is organized as follows: Section 2 talks about the related work. Section 3 describes our proposed approach in detail. Section 4 presents the experiments that support our approach. Finally we conclude our paper and present the future work in Section 5.

2   Related Work

The work proposed in [4] uses Freeling tool, common NE recognizer for English and Spanish to identify the NEs present in both the languages. But, it requires languages involved in the corpora to be of the same linguistic family. Such facility is not available for the Indian languages since they don't belong to a common linguistic family. Work proposed in [5] performed linguistic analysis such as lemmatization, morphological analysis to recognize the NEs present in the data. They represented each document with keywords and the extracted NEs and performed a Shared Nearest Neighbor (SNN) clustering algorithm for forming final clusters. Friburger *et al.* [6] have created their own Named Entity extraction tool based on a linguistic description with automata. The tool uses finite state transducers, which depends on the grammar of proper names. Authors in [7] have used the aligned English-Italian WordNet predicates present in MultiWordNet [8] for Multilingual named entity recognition. In all the above systems discussed, the authors used language dependent resources or tools to extract the NEs present in the data. Hence, such systems face the problem of extendability of their approches.

The work proposed in [9] did not make use of any non-English linguistic resources or tools such as WordNet or POS tagger. Instead, they used Wikipedia structure (Category, Interwiki links, etc.) to extract the NEs from the languages. The expectation in this paper is that for any language in which Wikipedia is sufficiently well-developed, a usable set of training data needs to be obtained. Clearly, the Wikipedia coverage of under-resourced languages falls short of this requirement. Hence, we propose a completely language-independent approach to extract the NEs present in under-resourced Indian languages (Hindi and Marathi) by utilizing the NEs present in English (a high resourced language). The detailed description of the proposed approach is given in Section 3.

3   Proposed Approach

In this section, we detail the two phases involved in the proposed approach. Phase-1 involves identification of the NEs present in Hindi and Marathi languages. These NEs are later utilized in Phase-2 for the formation of multilingual clusters.

3.1   Phase-1: NE Identification

As mentioned earlier, NEs such as persons, locations, organizations play a major role in measuring the document similarity. All such NEs present in English documents are identified using the Stanford Named Entity Recognizer[2](Stanford NER). As a pre-processing step, all the English documents present in the dataset are processed using MontyLemmatiser[3] to obtain the corresponding lemma forms. All the English, Hindi and Marathi text documents are then represented using the classical vector space model [10]. It represents the documents as a vector of keyword based features following the "bag of words" notation having no ordering information. The values in the vector are TFIDF scores. Instead of maintaining a seperate stopword list for every language, any word that appears in more

---

[2] http://nlp.stanford.edu/ner/index.shtml
[3] http://web.media.mit.edu/~hugo/montylingua

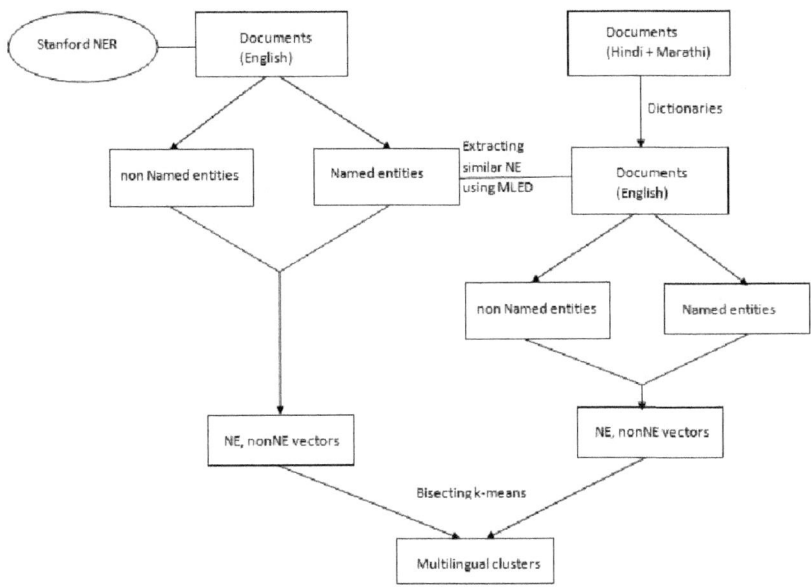

Fig. 1. MDC based on Named entities

than 60% of the documents is considered as a stopword. Hindi and Marathi documents are mapped onto a common ground representation (English) using various dictionaries. Experiments are conducted based on the usage of different dictionaries such as Wiki dictionary [2], bilingual dictionaries (Shabanjali dictionary[4] and Marathi-Hindi[5] dictionary). The translated versions of non-English (Hindi, Marathi) documents are also converted into base forms using MontyLemmatiser.

In order to identify the NEs present in non-English documents, the NEs present in all English documents are utilized. All the non-English words after being translated into English are compared with the NEs in English documents and words which have an exact match are identified as the NEs of corresponding non-English documents. After identifying the NEs in all non-English documents, a NE separator function is used to represent each document in the dataset with two vectors namely a NE vector and a nonNE vector. The NE vector contains only NEs present in the document. Whereas, the nonNE vector contains the remaining words of that document. In both these vectors the values are TFIDF scores.

3.2 Phase-2: Multilingual Document Clustering Based on Named Entities (MDC$_{NE}$)

Document clustering is an automatic grouping of text documents into clusters so that documents within a cluster have high similarity in comparison to one

---

[4] http://ltrc.iiit.net/onlineServices/Dictionaries/Dict_Frame.html
[5] http://ltrc.iiit.net/onlineServices/Dictionaries/Dict_Frame.html

another, but are dissimilar to documents in other clusters. Various clustering approaches (such as Hierarchical clustering, Partitioning clustering, etc.) are available for cluster formation. Steinbach *et al.* [11] compared different clustering algorithms and concluded that Bisecting k-means performs better than the standard k-means and agglomerative hierarchical clustering. We used Bisecting k-means algorithm for the cluster formation as it combines the strengths of partitional and hierarchical clustering methods by iteratively splitting the biggest cluster using the basic k-means algorithm. As mentioned in the previous section, each document is represented with two vectors namely NE vector and nonNE vector. In the proposed approach (MDC$_{NE}$), these vectors are linearly combined in measuring the document similarities using Eq. (1) and clustering is performed using Bisecting k-means algorithm. For the evaluation of Bisecting k-means algorithm, we have experimented with fifteen random k values between 30-70 and the average F-score, Purity and NMI values are considered as final clustering results.

The similarity between two documents is calculated by linearly combining the corresponding NE and nonNE vectors. We choose the cosine distance to measure the similarity of two documents (d$_i$ and d$_j$) which is defined as:

$$sim(d_i, d_j) = \alpha * (\frac{dim_1}{dim_h}) * sim^{NE} + \beta * (\frac{dim_2}{dim_h}) * sim^{nonNE} \qquad (1)$$

where dim_1 is the dimension of the NE vector, dim_2 is the dimension of the nonNE vector and $dim_h \in \max\{dim_1, dim_2\}$. The sim value is calculated as:

$$sim = cos(v_i, v_j) = \frac{v_i.v_j}{|v_i| * |v_j|} \qquad (2)$$

where vectors v_i, v_j belongs to either NE vector or nonNE vector of documents d_i and d_j respectively. The coefficients α and β indicate the importance of NE vector and nonNE vector respectively ($\alpha + \beta = 1$) in measuring the document similarities. In section 4.1, we present the evaluation criteria to determine the α and β values.

4   Experimental Evaluation

We have conducted experiments using the FIRE 2010 dataset available for the ad-hoc cross lingual document retrieval task. The data consists of news documents collected from 2004 to 2007 for each of the English, Hindi, Bengali and Marathi languages from regional news sources. There are 50 query topics represented in each of these languages. We have considered English, Hindi and Marathi documents for our experiments. We used the topic-annotated documents in English, Hindi and Marathi to build clusters. To introduce noise, we have added topic irrelevant documents that constitue 10 percent of topic documents. Some topics are represented by 8 or 9 documents whereas others are represented by about 50 documents. There are 2182 documents in the resulting

Table 1. Clustering schemes based on different combinations of vectors

Evaluation measure	System-1 Bilingual dictionaries		System-2 Bilingual + Wiki dictionaries		System-3 Wiki dictionary	
	$MDC_{Keyword}$ (*baseline*)	MDC_{NE}	$MDC_{Keyword}$	MDC_{NE}	$MDC_{Keyword}$	MDC_{NE}
F-Score	0.553	0.613	0.619	0.660	0.662	**0.710**
Purity	0.657	0.692	0.687	0.720	0.737	**0.762**
NMI	0.712	0.743	0725	0.752	0.761	**0.793**

collection of which, 650 are in English, 913 are in Hindi and 619 in Marathi. Cluster quality is evaluated by F-score [11], Purity [12] and NMI [13] measures.

F-score combines the information of precision and recall. To compute Purity, each cluster is assigned to the class which is most frequent in the cluster, and then the accuracy of this assignment is measured by counting the number of correctly assigned documents and dividing by total number of documents (N). High purity is easy to achieve when the number of clusters is large - in particular, purity is 1 if each document gets its own cluster. Thus, we cannot use purity to trade off the quality of the clustering against the number of clusters. A measure that allows us to make this tradeoff is Normalized Mutual Information or NMI.

$$NMI(\Omega, C) = \frac{I(\Omega, C)}{[H(\Omega) + H(C)]/2} \tag{3}$$

where I is the Mutual Information and H is entropy of the system.

$$I(\Omega, C) = \sum_k \sum_j \frac{P(\omega_k \cap c_j)}{N} \log \frac{N * P(\omega_k \cap c_j)}{P(\omega_k) * P(c_j)} \tag{4}$$

$$H(\Omega) = -\sum_k P(\omega_k) \log P(\omega_k) \tag{5}$$

where $P(\omega_k)$, $P(c_j)$, and $P(\omega_k \cap c_j)$ are the probabilities of a document being in cluster ω_k, class c_j, and in the intersection of ω_k and c_j, respectively.

The accuracy of the NE identification System of Phase-1 is determined using the following measures:

$$NE_{Precision} = \frac{NEs_{correctlyIdentified}}{NEs_{totalNEsIdentified}}. \tag{6}$$

$$NE_{Recall} = \frac{NEs_{correctlyIdentified}}{NEs_{totalNEsPresent}}. \tag{7}$$

4.1   Discussion

For the evaluation of NE identification system, we have randomly selected 90 documents from Hindi and Marathi dataset. Three experts from the linguistics department are given 30 documents each to manually identify the NEs present in those documents. The accuracy of NE identification system is determined using

Eq. (6) and Eq. (7) and the results obtained are shown in Table-2. For evaluation of the Bisecting k-means algorithm, we have experimented with fifteen random k values between 30-70 and the average F-score, Purity and NMI values are considered as the final clustering results. Three systems are formed based on the usage of different dictionaries such as Wiki dictionary, bilingual dictionaries (Shabanjali dictionary and Marathi-Hindi dictionary) in cluster formation. Linear combinations of NE vectors and nonNE vectors are examined for calculating document similarities in cluster formation.

In System-1, Hindi and Marathi documents are translated to English using only bilingual dictionaries and clustering is performed based on keywords, which is considered as the *baseline* for our experiments. In System-2, both Wiki dictionary and bilingual dictionaries are combinedly used to translate Hindi and Marathi documents into English. Whereas in System-3, only Wiki dictionary is used for the translation. The results obtained in System-1, System-2 and System-3 are shown in Table-1. Results obtained in System-2 shows that clustering based on keywords using bilingual dictionaries and Wiki dictionary together has yielded better results over baseline. Whereas the results obtained in System-3, which is a language-independent approach, shows that clustering based on keywords using only Wiki dictionary has yielded better results over System-2 and the baseline. This might be due to the fact that using bilingual dictionaries create the problem of word sense disambiguation, whereas in Wiki dictionary this problem is evaded as titles of wikipedia are aligned only once. In all the three systems clustering based on NEs has yielded better results over clustering based on keywords which shows the importance of the proposed approach. The reason for improvement in the results might be due to the fact that Wiki dictionary covers broader set of NEs which play a major role in clustering. In all these three systems the α value is determined in training phase, details of which are explained below.

Training Phase. Training data constitutes around 60% (1320 documents) of the total documents in the dataset. In these 1320 documents, 400 documents are in English, 550 are in Hindi and 370 in Marathi. The α value is determined by conducting experiments on the training data using Eq. (1). Bisecting k-means algorithm is performed on the training data by varying the α values from 0.0 to 1.0 with 0.1 increment ($\beta = 1-\alpha$). Finally, α is set to the value for which best cluster results are obtained. In our experiments, it is found that setting α value to 0.8 and β to 0.2 has yielded good results in System-1 and System-3. Whereas setting α to 0.7 and β to 0.3 has yielded good results in System-2.

Testing Phase. Test data constitutes around 40% (862 documents) of the total documents in the dataset. Out of these 862 documents, 250 documents are in English, 363 are in Hindi and 249 in Marathi. In all the three systems, Bisecting k-means algorithm is performed on the test data, after setting the α and β values obtained in training phase, using Eq. (1) in similarity calculation.

5   Conclusion and Future Work

In this paper we proposed an approach to identify the NEs present in under resourced languages by utilizing the NEs present in English. Bisecting k-means algorithm is performed for clustering multilingual documents based on the identified NEs. The results showcase the effectiveness of the NEs in clustering multilingual documents. From the results it can be concluded that NEs alone are not sufficient for forming better clusters. NEs when combined along with the nonNEs have yielded better clustering results. Our approach is completely language-independent as we haven't used any non-English linguistic resources (such as lemmatizers, NERs, etc.) for processing Hindi and Marathi documents. Instead, we have created alternatives such as Wiki dictionary (built from Wikipedia) and MLED, which is a replacement for lemmatizers. The proposed approach is easy to re-implement and especially useful for the under-resourced languages where the availability of the tools and resources such as dictionaries, lemmatizers, Named Entity Recognizer, etc., is a major problem.

We plan to extend the proposed approach which implements only static clustering to handle the dynamic clustering of multilingual documents. Also, we would like to consider comparable corpora of different languages to study the applicability of our approach.

References

1. Pirkola, A., Hedlund, T., Keskustalo, H., Järvelin, K.: Dictionary-based cross-language information retrieval: Problems, methods, and research findings. Information Retrieval 4, 209–230 (2001)
2. Kumar, N.K., Santosh, G., Varma, V.: Multilingual document clustering using wikipedia as external knowledge. In: Proceedings of IRFC (2011)
3. Santosh, G., Kumar, N.K., Varma, V.: Ranking multilingual documents using minimal language dependent resources. In: Proceedings of 12th International Conference on Intelligent Text Processing and Computational Linguistics, Tokyo, Japan,
4. Montalvo, S., Martínez, R., Casillas, A., Fresno, V.: Multilingual document clustering: an heuristic approach based on cognate named entities. In: Proceedings of the 21st International Conference on Computational Linguistics and the 44th Annual Meeting of the Association for Computational Linguistics (ACL), pp. 1145–1152. Association for Computational Linguistics, Morristown (2006)
5. Romaric, B.M., Mathieu, B., Besançon, R., Fluhr, C.: Multilingual document clusters discovery. In: RIAO, pp. 1–10 (2004)
6. Friburger, N., Maurel, D., Giacometti, A.: Textual similarity based on proper names. In: Proceedings of the workshop Mathematical/Formal Methods in Information Retrieval (MFIR 2002) at the 25 th ACM SIGIR Conference, pp. 155–167 (2002)
7. Negri, M., Magnini, B.: Using wordnet predicates for multilingual named entity recognition. In: Proceedings of The Second Global Wordnet Conference, pp. 169–174 (2004)
8. Pianta, E., Bentivogli, L., Girardi, C.: Multiwordnet: Developing an aligned multilingual database. In: Proceedings of the 1st International Global WordNet Conference, Mysore, India (2002)

9. Richman, A.E., Schone, P.: Mining wiki resources for multilingual named entity recognition. In: Proceedings of ACL 2008 HLT (2008)
10. Salton, G., Wong, A., Yang, C.S.: A vector space model for automatic indexing. Commun. ACM 18, 613–620 (1975)
11. Steinbach, M., Karypis, G., Kumar, V.: A comparison of document clustering techniques. In: TextMining Workshop, KDD (2000)
12. Zhao, Y., Karypis, G.: Criterion functions for document clustering: Experiments and analysis. Technical report, Department of Computer Science, University of Minnesota. (2002)
13. Zhong, S., Ghosh, J.: Generative model-based document clustering: a comparative study. Knowledge and Information Systems 8, 374–384 (2005)

Multilingual Question-Answering System in Biomedical Domain on the Web: An Evaluation

María-Dolores Olvera-Lobo[1,2] and Juncal Gutiérrez-Artacho[2]

[1] CSIC, Unidad Asociada Grupo SCImago, Madrid, Spain
[2] University of Granada, Spain
{molvera,juncalgutierrez}@ugr.es

Abstract. Question-answering systems (QAS) are presented as an alternative to traditional systems of information retrieval, intended to offer precise responses to factual questions. An analysis has been made of the results offered by the QA multilingual biomedical system HONqa, available on the Web. The study has used a set of 120 biomedical definitional questions (*What is...?*), taken from the medical website WebMD, which were formulated in English, French, and Italian. The answers have been analysed using a serie of specific measures (MRR, TRR, FHS, precision, MAP).

The study confirms that for all the languages analysed the functioning effectiveness needs to be improved, although in the multilingual context analysed the questions in the English language achieve better results for retrieving definitional information than in French and Italian.

Keywords: Multilingual information, Multilingual Question Answering Systems, Restricted-domain Question Answering Systems, HONqa, Biomedical information, Evaluation measures.

1 Introduction

The advent of the Web and its subsequent expansion has provided the general public with access to enormous volumes of information, offering unquestionable benefits. Nevertheless, this has also brought disadvantages such as overloads of information — which in this environment is even more acute— or the fact that much of the information is incorrect, incomplete, or inaccurate, whether intentionally so or not. Consequently, it becomes indispensable to develop tools and procedures that enable the user to acquire reliable information that is relevant for a particular consultation. This is the challenge that faces Information Retrieval (hereafter IR). Some of the efforts to improve IR in the Web have focused on the design and development of question-answering systems (hereafter QAS).

This work evaluates the multilingual search for definitional responses in the context of restricted-domain QAS. For this, the results offered by the multilingual biomedical QAS HONqa, available on the Web, were analysed. A set of 120 responses (in English, French, and Italian) related to this thematic field were assessed by a series of measures applicable to such systems.

P. Forner et al. (Eds.): CLEF 2011, LNCS 6941, pp. 83–88, 2011.

2 Question Answering Systems: Beyond Information Retrieval

According to a study by Ely [1], medical specialists invest an average of more than two minutes searching for information related to questions that arise and, despite the time taken up, adequate answers are often not found. In this sense, several works have demonstrated the confidence of medical specialists in the use of QAS as a method of searching and retrieving specialized information [2-3]. Patients have also increasingly consulted these systems, before and after seeing the doctor, to gather information on the nature of the illness, treatment recommendations, contraindications, etc. [4]

QAS are designed to offer understandable responses to factual questions of specialized content rapidly and precisely in such a way that the user does not have to read the complete documents to satisfy a particular query. These systems begin with the user's question in order to construct coherent answers in everyday language [5]. The functioning of the QAS is based on short-answer models [6], since the potentially correct answer does not go beyond a number, a noun, a short phrase, or a brief fragment of text. Then the QAS locates and extracts one or several responses from different sources, according to the subject of the consultation [7]. Finally an evaluation is made and information that is redundant or that does not answer the question properly is eliminated in order to present specific responses designed to satisfy the information needed by the user [8-10].

For the environment of the Web, some of QAS have been developed only as prototypes or demos, and are very rare in systems available to the final user. Nonetheless, some QAS of a general domain can currently be consulted, as they are capable of addressing questions on very diverse subjects (such as START) and from very specific domains (such as EAGLi), which focus on a given context. In addition, the QA systems try to overcome the limitations of the traditional tools of information retrieval, such as the consultations being monolingual.

The appropriate retrieval tools that enable the procedure known as cross-lingual information in retrieval (CLIR) would enable consultations in several languages with information retrieval in all the languages accepted by the system [11]. Although the Cross-lingual QAS of restricted domain are not yet avalaible for the final users, on the Web it begins to find some on the sphere of multilingual QAS (such as HONqa).

3 Method Section

The methodology applied used 120 biomedical questions concerning definitions on diverse medical subjects. The questions were formulated as consultations of the type "What is" in the search engine of the website WebMD, created in the USA by medical specialists to resolve doubts held by patients. From the questions that this portal provides, a set was selected to be translated by a team of professional translators of French and Italian, and from this initial set, 120 questions that elicited responses in the three languages in the system were selected. These constituted the body of the questions used. The main biomedical aspects related to the selected questions were diseases, operations, treatments, syndromes, and symptoms. The set of questions used passed the validity test with a Chronbach's alpha of 0.936. HONqa, the QAS evaluated in this work, was developed by the *Health On the Net*

Foundation. It is a multilingual system that retrieves information in English, French, and Italian [12].

The responses offered by the system were evaluated by a group of experts from different medical areas, as incorrect, inexact, or correct, according to the evaluation methodology proposed in CLEF [13]. Questions that were answered properly and did not add irrelevant information were considered correct. All the answers that resolved the question but added irrelevant information were considered inexact. Finally, answers that contained irrelevant information with regard to the question were considered incorrect. From the evaluation of the answers retrieved, the evaluation measures were applied are: *Mean Reciprocal Rank* (MRR), which assigns the inverse value of the position in which the correct answer is found, or zero if there is no correct response; *Total Reciprocal Rank* (TRR), useful for evaluating the existence of several correct responses offered by a system to the same query; *First Hit Success* (FHS, which assigns a value of 1 if the first answer offered is correct, and a value of 0 if it is not; *Precision,* which measures the ratio of retrieval responses are relevant to the query; and *Mean Average Precision* (MAP), which measures the average precision of a set of queries for which the answers are arranged by relevance.

The users access to the QA systems for their quickness and precision, and they are not likely to read a long list of answers for each question. So their futility point [14-16] –the maximum number of responses they would be willing to begin browsing through–, will be probably more exigent than others Information Retrieval Systems. For this reason, only the first five answers in each of these systems were analysed – although the mean of the answers retrieved by the system in the three languages approached and in some cases exceeded.

4   Results Section

The average of the total answers retrieved by the system was 47.46 in the case of English, 27.36 for French, and 25.03 for Italian.

Table 1. Answers retrieved by HONqa in the three languages

	Total answers	Average of Answers	Answers analyzed	Correct answers	Inexact answers	Incorrect answers
English	5695	47.46	589	287 (48.73%)	67 (11.4%)	235 (39.9%)
French	3283	27.36	573	52 (9.07%)	124 (21.6%)	397 (69.3%)
Italian	3123	25.03	585	32 (5.47%)	95 (11.6%)	458 (82.9%)

The volume of answers retrieved in English was substantially higher (5695 answers retrieved) than in the other cases, the other two languages registering similar values (3283 for French and 3123 for Italian).

The correct answers were present in greater measure in the English version of the system, which properly responded to more than 48% of the cases, whereas French offered a low rate of 9.07% and Italian provided only 2.05%. The number of imprecise answers was higher in French (21.64%), followed by Italian (16.24%). In relation to the incorrect answers, the number was very high in all three languages, exceeding 50% of the total in French (397) and Italian (458).

This behaviour directly influenced the results found when applying the evaluation measures proposed. The MRR value for the responses offered in the three languages reflect the above comments. While the results of the English option were quite plausible, at 0.76, the other two languages offered very poor results (0.19 for French and 0.13 for Italian), indicating the low reliability of the first response by the system for these languages.

In relation to the TRR measure, which considers all the answers correct among the first 5 results analysed, it was found that, except for English the results did not substantially improve. FHS is an important measure, as the users often tend to focus on the first response retrieved, skipping the rest. It was found that more than 50% of the answers offered in English (0.575) provided an initial correct answer while the other cases were not encouraging (0.12 in French and 0.06 in Italian). MAP is a widely used measure that offers an overall idea of the functioning of the system. The evaluation of the system did not register an adequate level for any of the languages analysed.

Table 2. Evaluation measures (P=Precision, P*=Precision considering also the inexact answers, P@3=Precision of the 3 first results, P@3*=Precision of the 3 first results including inexact answers)

	MRR	TRR	FHS	P	P*	P@3	P@3*	MAP
English	0.76	1.55	0.575	0.55	0.65	0.57	0.67	0.25
French	0.19	0.27	0.12	0.10	0.31	0.11	0.32	0.05
Italian	0.13	0.15	0.06	0.05	0.16	0.06	0.15	0.03

The precision value is closely related to the rest of the measures discussed above. The precision was measured, on the one hand, considering as relevant only the responses scored as correct (measures P and P@3) and, on the other hand, considering also the imprecise answers (measures P* and P@3*) as relevant –that is, being more flexible to evaluate an answer as adequate. In this latter case, clearly, the precision values significantly increased in some cases. Nevertheless, as with the rest of the measures, there was a marked different between English and the other languages. On considering P@3 or only the precision of the first three results, the values found for this new measure only weakly improved though not very different from the previous values. This indicates that the arrangement of the answers retrieved according to their relevance to the question was not the best. The small number of correct answers in some cases made the recall values of the QAS very low, except in the case of English.

5   Conclusions

The analysis of the results from posing 120 questions in the QA system of the biomedical domain HONqa has enabled the evaluation of its functioning in the retrieval of multilingual information by applying specific measures and analysing the information sources used for each language. Despite the restrictions that these systems show, the study indicates that this QA system is valid and useful for the retrieval of definitional medical information, mainly in the English language, although it is not yet the most advisable resource to gather multilingual information in a quick and precise way.

The search for multilingual responses in the context of the Web still needs to progress a long way to reach the effectiveness levels of general retrieval systems, and especially in monolingual ones. Nevertheless, the results are promising as they show this type of tool to be a new possibility within the sphere of precise, reliable, and specific information retrieval in a brief period of time.

References

1. Ely, J.W., Osheroff, P.N., Ebell, M., Bergus, G., Barcey, L., Chambliss, M., Evans, E.: Analysis of questions asked by family doctors regarding patient care. British Medical Journal 319, 358–361 (1999)
2. Lee, M., Cimino, J., Zhu, H.R., Sable, C., Shanker, V., Ely, J., Yu, H.: Beyond Information Retrieval –Medical Question Answering. AMIA, Washington DC (2006)
3. Yu, H., Kaufman, D.: A cognitive evaluation of four online search engines for answering definitional questions posed by physicians. In: Pacific Symposium on Biocomputing, vol. 12, pp. 328–339 (2007)
4. Zweigenbaum, P.: Question answering in biomedicine. In: Rijke, Webber (eds.) Proceedings Workshop on Natural Language Processing for Question Answering, EACL 2003, pp. 1–4. ACL, Budapest (2005)
5. Costa L.F., Santos, D.: Question Answering Systems: a partial answer (SINTEF, Oslo) (2007)
6. Blair-Goldensohn, S., McKeown, K., Schlaikjer, A.H.: Answering Definitional Questions: A Hybrid Approach. New Directions in Question Answering 4, 47–58 (2004)
7. Olvera-Lobo, M.D., Gutiérrez-Artacho, J.: Language resources used in Multi-lingual Question Answering Systems. Online Information Review 35(4) (forthcoming, 2011)
8. Cui, H., Kan, M.Y., Chua, T.S., Xiao, J.: A Comparative Study on Sentence Retrieval for Definitional Question Answering. In: SIGIR Workshop on Information retrieval for Question Answering (IR4QA), Sheffield (2004)
9. Tsur, O.: Definitional Question-Answering Using Trainable Text Classifiers. PhD Thesis. University of Amsterdam (2003)
10. Olvera-Lobo, M.D., Gutiérrez-Artacho, J.: Question-Answering Systems as Efficient Sources of Terminological Information: Evaluation. Health Information and Library Journal 27(4), 268–274 (2010)
11. Diekema, A. R.: Translation Events in Cross-Language Information Retrieval: Lexical ambiguity, lexical holes, vocabulary mismatch, and correct translations. PhD Thesis. University of Syracuse (2003)
12. Cruchet, S., Gaudinat, A., Rindflesch, T., Boyer, C.: What about trust in the Question Answering world? In: AMIA 2009 Annual Symposium, San Francisco (2009)
13. Blair, D.C.: Searching biases in large interactive document retrieval systems. Journal of the American Society for Information Science 31(4), 271–277 (1980)

14. Peters, C.: What Happened in CLEF 2009: Introduction to the Working Notes. In: Peters, C., Di Nunzio, G.M., Kurimo, M., Mostefa, D., Penas, A., Roda, G. (eds.) CLEF 2009. LNCS, vol. 6241, pp. 1–12. Springer, Heidelberg (2010),
    http://www.clefcampaign.org/2009/working_notes/
    CLEF2009-intro.pdf
15. Raved, D.R., Qi, H., Wu, H. Fan, W.: Evaluating Web-based Question Answering Systems. Technical Report, University of Michigan (2001)
16. Salton, G., Mc Gill, M.J.: Introduction to Modern Information Retrieval. Mc Graw-Hill, New York (1983)

Simulation of Within-Session Query Variations Using a Text Segmentation Approach

Debasis Ganguly, Johannes Leveling, and Gareth J.F. Jones

CNGL, School of Computing, Dublin City University, Dublin-9, Ireland
{dganguly,jleveling,gjones}@computing.dcu.ie

Abstract. We propose a generative model for automatic query reformulations from an initial query using the underlying subtopic structure of top ranked retrieved documents. We address three types of query reformulations: a) *specialization*; b) *generalization*; and c) *drift*. To test our model we generate three reformulation variants starting with selected fields from the TREC-8 topics as the initial queries. We use manual judgements from multiple assessors to measure the accuracy of the reformulated query variants and observe accuracies of 65%, 82% and 69% respectively for specialization, generalization and drift reformulations.

1 Introduction

Laboratory information retrieval (IR) evaluation has generally focused on single query search where a query is applied to an IR system and the effectiveness of retrieving relevant documents is evaluated. However, it is commonly observed that multiple versions of a query are iteratively modified and applied to retrieval systems [1]. Three patterns of query reformulation as observed in real-life search behaviour [2] are the following: a) specialization, where the reformulated query expresses a more specialized information need as compared to the initial query; b) generalization, where the refined information need is more general and covers a broader scope in comparison to the initial query; c) drift (parallel reformulation), where the reformulated query drifts away to another aspect of the initial information need instead of moving to more general or more specific needs.

The topic collections of standard ad hoc tracks at IR evaluation workshops provide a set of unrelated topics and hence fail to evaluate the performance of a retrieval system over a session of related queries. The Session Track of TREC [3] aims to evaluate retrieval systems over an entire session of user queries rather than on separate independent topics. The topic creation phase of the Session Track 2010 involved starting with Web-track diversity topics, comprising of separate sub-topics representing different facets of information need, sampled from the query logs of a commercial search engine.

This paper describes a model to simulate user interactions over a browsing session by automatically generating the above mentioned three types of reformulated queries. We also look at ways of characterizing the reformulations based on the differences in retrieved sets of documents between the initial and the

P. Forner et al. (Eds.): CLEF 2011, LNCS 6941, pp. 89–94, 2011.

reformulated queries. The reformulations are qualitatively evaluated by manual judgments. Related work on automatic query reformulation for web-search to better answer the initial original information need itself, includes that of Dang and Croft [4] which uses anchor text to reformulate a query by substituting some of the original terms. Wang and Zhai [5] exploited co-occurrence of terms in search-engine query logs to add terms to correct the mis-specification and under-specification of a query. Our work is different in the sense that we do not intend to improve the initial query; but rather seek to move the query towards a more specific subtopic or a broader topic which is thus associated with a change of the information need. We also aim to develop topic variants on a large scale for ad hoc retrieval collections which do not possess meta-information such as query logs or anchor texts.

The rest of the paper is organized as follows: Section 2 discusses our methodology for generating query sessions, Section 3 hypothesizes the expected characteristics of the retrieved set of documents for the three reformulated versions, Section 4 describes the experiments performed, and finally Section 5 concludes the paper with directions for future work.

2   Automatically Generating Query Reformulations

A real user will often start with a general query at the beginning of a search session, since initially he may not be aware of the more specific aspects of his information need [6]. Then as he views retrieved documents his knowledge about the topic increases and he may get interested in more specific details about the topic. His reformulation of the initial information need is thus based on the contents of the more focused subtopics. If the user is interested in a more specific reformulation, he is likely to choose terms which occur frequently in one or a few subtopics. Whereas if he starts with a specific query, he may become interested in a more general formulation, in which case he would choose terms which are not concentrated in one of the subtopics, but occur abundantly throughout the document.

Our earlier work [7] shows that a text segmentation based approach can be used to simulate the reformulation patterns of real users and automatically generate the query variants. The differences with our earlier work are as follows: i) in contrast to an additive model for generating general reformulations, we employ removal of words; ii) we propose a method for generating the parallel reformulations; iii) the algorithm for generating the specific reformulations is slightly different in the sense that term selection scores are accumulated over top ranked documents. We now look at each of the reformulation types in more detail.

Text Segmentation or Text-Tiling [8] is the process of decomposing a text into blocks of coherent textual content called segments. Thus each segment content is particularly focused on one subtopic. Our generative model tries to utilize the fact that a term indicative of a more specific aspect of an initial information need, typically is densely distributed in the textual contents of a particular segment. We use C99 [9], which is shown to perform better than Hearst's original

text-tiling algorithm, for segmenting documents. To characterize specific reformulation terms, we assign scores to terms considering the following two factors:
a) how frequently a term t occurs in a segment s, denoted by $\mathtt{tf}(t,s)$, and how
exclusive the occurrence of t in s is, as compared to other segments of the same
document, denoted by $\frac{|S|}{\mathtt{sf}(t)}$, where $|S|$ is the number of segments in that doc-
ument and $\mathtt{sf}(t)$ is the number of segments in which t occurs; b) how rare the
term is in the entire collection, measured by the document frequency ($\mathtt{df}$), the
assumption being that rare terms are more likely to be specific terms. For specific reformulations, we compute the term scores for the most similar segment to
the query assuming that this is precisely the section which "catches the eye" of
a real-life reader and that adding terms from this segment can potentially shift
the original query to a more specific information need.

$$\phi(t,s) = a \cdot \mathtt{tf}(t,s)\frac{|S|}{\mathtt{sf}(t)} + (1-a) \cdot \log\frac{|D|}{\mathtt{df}(t)} \tag{1}$$

$$\psi(t,d) = a \cdot \mathtt{tf}(t,d)\frac{\mathtt{sf}(t)}{|S|} + (1-a) \cdot \log\frac{|D|}{\mathtt{df}(t)} \tag{2}$$

We use a mixture model to calculate term scores, as shown in Equations 1 and 2.
Equation 1 assigns higher values to terms which occur frequently in a segment,
occur only in a few segments, and occur infrequently in the collection. The
working steps of the proposed method to generate query variants are as follows:

1. For each top ranked R documents do Steps 2-5.
2. Segment a document d into $\{s_1, s_2, \dots s_n\}$ by executing C99.
3. Let s_{sel} be the segment with maximum and minimum number of matching query
   terms respectively for specific and parallel reformulations.
4. For general reformulation, score each original query term t by $\psi(t,d)$; otherwise
   score each term $t \in s_{sel}$ by $\phi(t, s_{sel})$ for specific and parallel reformulations.
5. Average the term scores over documents.
6. Sort each term of the table by its score and add(substitute) top n new terms to
   the original query for specific(parallel) reformulation type. Retain the top n terms
   in the query removing the rest for a general reformulation.

When we are done processing a document, we store the term scores from this
document and move on to extract terms from the next document, merging the
new scores with the previously stored term scores. The state of the stored terms
is useful in simulating a user who keeps track of the more specific sub-topical
terms as he keeps on reading documents retrieved in response to the initial query.
Although he reads each document in turn, his decision of which terms to add for
reformulating the query is a global one based on the information gained from
all the top ranked documents read. Merging term scores by averaging out the
previous score of a term with the score of that term in the current document
simulates this behaviour.

In contrast to a more specific term, a more general term is distributed uni-
formly throughout the entire document text [8]. So an obvious choice is to score
a term based on the combination of term frequency in the whole document (instead of frequency in individual segments) and segment frequency (instead of

inverse segment frequency), where $\mathtt{tf}(t, d)$ is the number of occurrences of t in d (see Equation 2). The model used in generalization removes terms in contrast to the additive model for the specific reformulation type. More precisely, it involves removal of terms of higher $\phi(t, s)$ in the initial query with those having lower ones, thus making general reformulation an inverse to specialization.

To simulate a parallel reformulation, we assume that the user reformulates the initial query by adding specific terms from the least similar (to the initial query) subtopics of the read documents. The differences with the specialization reformulation are that firstly term scores are computed from the least similar segment in contrast to the most similar one and secondly none of the initial query terms are retained in the reformulated query, thus resulting in a substitution based model where we throw away all the initial query terms and add n new terms. Although the reformulated query does not share any common terms, it expresses a parallel information need based on the contents of the documents being read which is different from starting a new session.

3   Reformulation Effect on Retrieval

The original query terms ought to be semantically related to the added specialization terms to make the reformulation precise whereas the reverse is true for generalization reformulations. Although in this paper we have looked at generating the reformulations from an IR perspective, simulating the behaviour of a real-life searcher, an alternative collection independent way of generating reformulations would be to use a thesaurus such as the WordNet which encodes the hierarchic relationships between words. Thus, specialization can be achieved by addition of hyponyms or meronyms, whereas generalization can involve addition of hypernyms or holonyms or when viewed as the inverse of specialization - removal of hyponyms or meronyms. It is expected that there will be a smaller number of documents in the collection pertaining to a more specific information need. As the query becomes more specific in nature, the top documents retrieved for the initial query become more general to the new information need and shift down the ranked list. Thus, if we measure overlap of the two ranked lists at specific cut-off points, we would expect a low overlap in the top 10 or top 20, whereas a high degree of overlap beyond that. To measure the shift in the rankings of top ranked documents, we define the *net perturbation* of top m documents for a query q as:

$$p(q, m) = \frac{1}{m} \sum_{k=1}^{m} \delta_k \tag{3}$$

where $\delta_k = |(\mathtt{newrank}(d_k) - k|$ if d_k exists in the ranked list of the reformulated query, and $\delta_k = 1000$ otherwise. For the specific reformulations we would expect a lower net perturbation value as compared to generalization and drift. Since generalization involves removal of terms from the original query it opens up a wider range of documents to be retrieved for the reformulated query, for which we thus expect a lower degree of overlap of the two ranked lists. Since parallel reformulation often involves using substituted words, we expect further lower overlap and higher net perturbation of top ranked documents.

4 Experiments

We start with the titles of the TREC-8 topics as initial queries for generating the specific and parallel reformulation variants. For the general reformulations we use the description field of the TREC-8 topics as the initial query. We use 5 top ranked documents for reformulating each query with $a = 0.5$ for computing the $\phi(t, s)$ and $\psi(t, d)$ scores i.e. equal importance is given to the term distribution factor in a document and the *idf* factor of that term. We add at most 3 additional terms for the specific and parallel reformulations, and retain at most 2 terms from the description field to construct the general reformulations. Manual judgments regarding the quality of the generated queries were provided by two assessors through yes/no answers. Table 1 shows the assessor judgments on each reformulation type over 50 TREC-8 topics. From Table 1 we see that two assessors have the highest inter-agreement for the parallel reformulation type. This is expected because parallel reformulation always involves a change in information need and is more obvious to judge than the other two. The lowest inter-agreement is on specialization which can be attributed to the fact that such a reformulation involves addition of new words which ought to be semantically related to the original keywords, and the degree of semantic closeness is often subject to personal judgments. Table 1 also confirms the result set change hypothesis discussed in Section 3. Specific reformulations show the highest overlap with the initial retrieved set of documents. Drift reformulations exhibit minimum overlap. For general reformulations, overlap percentages are somewhere in between the two extremes. We also see that the specific and general reformulations are associated with an increase in overlap percentage with increasing cut-off rank, thus indicating that moving down the ranked list results in finding more documents already retrieved for the original query. However the drift reformulation exhibits a decrease in overlap with increasing cut-off rank, indicating that we find more unseen documents as we walk down the ranked list. Table 1 shows that the average net perturbation, as defined in Equation 3, is the least for specific reformulation type. Thus, the top 5 documents of the initial ranked list can be found not too far down the reformulated ranked list.

Table 1. Evaluation of the generated reformulations. $O(m)$ denotes the avg. % overlap and $p(m)$ denotes the avg. *net perturbation* of m top ranked docs.

Reformulation type	Manual assessments		Resultset measures				
	Assessor-1	Assessor-2	O(10)	O(20)	O(50)	O(500)	$p(5)$
Specific	39 (78%)	26 (52%)	39.0	38.1	42.7	44.7	367.9
General	39 (78%)	43 (86%)	22.4	22.5	24.5	32.2	2208.6
Parallel	34 (68%)	35 (70%)	12.0	10.2	8.6	5.89	3853.3

5   Conclusions

This paper presented a novel approach of generating simulated query sessions without using real-life search logs or external resources, simulating real-life reading behaviour by adding frequent sub-topical terms to form a more specific query, removing frequent non-uniformly distributed terms to form more general queries, and substituting dense sub-topical terms to construct parallel reformulations. Results show that our generative model can be used to produce query reformulations with an average accuracy of 65%, 82% and 69% for the specialization, generalization and drift reformulations respectively. This paper also provides arguments for the expected changes in the reformulated result set of documents in comparison with the initial retrieval set by introducing measures such as average percentage overlap at fixed number of documents and the average net perturbation. Retrieval results on the reformulated queries confirm the hypothesis put forward. Based on our findings we conclude that our automatic method of query generation can be used to generate topic variants on a large scale and thus create simulated user sessions with no manual intervention.

Acknowledgments. This research is supported by the Science Foundation Ireland (Grant 07/CE/I1142) as part of the Centre for Next Generation Localisation (CNGL).

References

1. Bates, Marcia, J.: The Design of Browsing and Berrypicking Techniques for the Online Search Interface. Online Review 13(5), 407–424 (1989)
2. Jansen, B.J., Booth, D.L., Spink, A.: Patterns of query reformulation during web searching. J. Am. Soc. Inf. Sci. Technol. 60, 1358–1371 (2009)
3. Kanoulas, E., Clough, P., Carterette, B., Sanderson, M.: Session track at TREC 2010. In: SIMINT Workshop SIGIR 2010. ACM, New York (2010)
4. Dang, V., Croft, B.W.: Query reformulation using anchor text. In: Proceedings of WSDM 2010, pp. 41–50. ACM, New York (2010)
5. Xuanhui, W., ChengXiang, Z.: Mining term association patterns from search logs for effective query reformulation. In: Proceedings of CIKM 2008, pp. 479–488 (2008)
6. Leveling, J., Ghorab, M.R., Magdy, W., Jones, G.J.F., Wade, V.: DCU-TCD@logCLEF 2010: Re-ranking document collections and query performance estimation. In: CLEF (Notebook Papers/LABs/Workshops) (2010)
7. Ganguly, D., Leveling, J., Jones, G.: Automatic generation of query sessions using text segmentation. In: SIR Workshop at ECIR 2011 (2011)
8. Hearst, M.: TextTiling: Segmenting text into multi-paragraph subtopic passages. CL 23(1), 33–64 (1997)
9. Choi, F.Y.Y.: Advances in domain independent linear text segmentation. In: Proceedings of the NAACL 2000, pp. 26–33 (2000)

Assessing the Scholarly Impact of ImageCLEF

Theodora Tsikrika, Alba G. Seco de Herrera, and Henning Müller

University of Applied Sciences Western Switzerland (HES–SO)
Sierre, Switzerland
theodora.tsikrika@acm.org, {alba.garcia,henning.mueller}@hevs.ch

Abstract. Systematic evaluation has an important place in information retrieval research starting with the Cranfield tests and currently with TREC (Text REtrieval Conference) and other evaluation campaigns. Such benchmarks are often mentioned to have an important impact in advancing a research field and making techniques comparable. Still, their exact impact is hard to measure. This paper aims at assessing the scholarly impact of the ImageCLEF image retrieval evaluation initiative. To this end, the papers in the proceedings published after each evaluation campaign and their citations are analysed using Scopus and Google Scholar. A significant impact of ImageCLEF could be shown through this bibliometric analysis. The differences between the employed analysis methods, each with its advantages and limitations, are also discussed.

1 Introduction

Evaluation campaigns in the field of information retrieval enable the reproducible and comparative evaluation of new approaches, algorithms, theories, and models, through the use of standardised resources and common evaluation methodologies within regular and systematic evaluation cycles. Over the years, several large–scale evaluation campaigns have been established at the international level, where major initiatives in the field of textual information retrieval include the Text REtrieval Conference[1] (TREC), the Cross–Language Evaluation Forum[2] (CLEF), the INitiative for the Evaluation of XML retrieval[3] (INEX), and the NTCIR Evaluation of Information Access Technologies[4]. Similar evaluation exercises are also carried out in the field of visual information retrieval, with TREC Video Retrieval Evaluation[5] (TRECVid), PASCAL Visual Object Classes challenge[6], MediaEval[7], and ImageCLEF[8] being among the most prominent.

These evaluation campaigns are predominantly based on the Cranfield paradigm [2] of experimentally assessing the worth and validity of new ideas in a laboratory setting through the use of *test collections*. Although this evaluation model has met with some criticism (see [12] for a discussion), such organised

---

[1] http://trec.nist.gov/
[2] http://www.clef-campaign.org/
[3] http://www.inex.otago.ac.nz/
[4] http://ntcir.nii.ac.jp/
[5] http://trecvid.nist.gov/
[6] http://pascallin.ecs.soton.ac.uk/challenges/VOC/
[7] http://www.multimediaeval.org/
[8] http://www.imageclef.org/

P. Forner et al. (Eds.): CLEF 2011, LNCS 6941, pp. 95–106, 2011.
© Springer-Verlag Berlin Heidelberg 2011

benchmarking activities have been widely credited with contributing tremendously to the advancement of information retrieval by (i) providing access to infrastructure and evaluation resources that support researchers in the development of new approaches, and (ii) encouraging collaboration and interaction between researchers both from academia and industry. Their contribution to the field is mainly indicated by the research that would otherwise not have been possible, i.e., research that heavily relies on the use of resources they provide. It is then reasonable to consider that their success can be measured to some extent by the scientific and possibly the economic impact of the research they foster.

The scientific impact of research is commonly measured by its scholarly impact, i.e., the publications derived from it and the citations they receive, and by additional indicators, such as filed patents, whereas its economic impact can be measured by the technology transfer efforts that result in commercial products and services. Other aspects, such as the scientific impact of the increased quality in evaluation methodologies or the economic impact of the time saved by researchers, who now reuse evaluation resources, rather than create them from scratch, are harder to assess. Recent investigations have reported on the scholarly impact of TRECVid [13] and on the economic impact of TREC [11]. Building on this work, this paper presents a preliminary study on assessing the scholarly impact of ImageCLEF, the cross–language image retrieval evaluation initiative that has been running as part of CLEF since 2003. To this end, it performs a citation analysis on a dataset of publications derived from ImageCLEF.

The rest of the paper is organised as follows: Section 2 provides a short introduction to ImageCLEF. Section 3 presents the bibliometric analysis method and tools, Section 4 the dataset of ImageCLEF publications considered, while Section 5 reports on the results of this analysis. Section 6 concludes this work.

2   The ImageCLEF Evaluation Campaign

ImageCLEF, the cross–language image retrieval annual evaluation campaign, was introduced in 2003 as part of CLEF and forms a natural extension to other CLEF tracks given the language neutrality of visual media. Motivated by the need to support multilingual users from a global community accessing the ever growing body of visual information, the main aims of ImageCLEF are: (i) to develop the necessary infrastructure for the evaluation of visual information retrieval systems operating in both monolingual and cross–language contexts, (ii) to provide reusable resources for such benchmarking purposes, and (iii) to promote the exchange of ideas towards the further advancement of the field of visual media analysis, indexing, classification, and retrieval.

To meet these objectives a number of tasks have been organised by Image-CLEF within two main domains: (i) medical image retrieval, and (ii) general (non–medical) image retrieval from historical archives, news photographic collections, and Wikipedia pages. These tasks can be broadly categorised as follows:

- *Ad hoc retrieval.* This simulates a classic document retrieval task: given a statement describing a user's information need, find as many relevant documents as possible and rank the results by relevance. In the case of cross–lingual retrieval, the language of the query is different from the language of the metadata used to describe the image. Ad hoc tasks have run since 2004 for medical retrieval and since 2003 for non–medical retrieval scenarios.

- *Object and concept recognition.* Although ad hoc retrieval is a core image retrieval task, a common precursor is to identify whether certain objects or concepts from a pre–defined set of classes are contained in an image (object class recognition), assign textual labels or descriptions to an image (automatic image annotation) or classify images into one or many classes (automatic image classification). Such tasks, including a medical image annotation and a robot vision task, have run since 2005.

- *Interactive image retrieval.* Since 2003, a user–centred task was run as a part of ImageCLEF and eventually followed by the interactive CLEF (iCLEF) track in 2005. Interaction in image retrieval can be studied with respect to how effectively the system supports users with query formulation, translation (for cross–lingual IR), document selection and examination.

Table 1 summarises the ImageCLEF tasks that ran between 2003 and 2010 and shows the number of participants for each task along with the distinct number of participants in each year. The number of participants and tasks offered by ImageCLEF has continued to grow steadily throughout the years, from four participants and one task in 2003, reaching its peak in 2009 with 65 participants and seven tasks. The number of participants, i.e., research groups that officially submit their results, is typically much smaller than the number of groups that register and gain access to the data; e.g., in 2010, 112 groups registered, but only 47 submitted results. Given its multi–disciplinary nature, ImageCLEF participants originate from a number of different research communities, including (visual) information retrieval, cross–lingual information retrieval, computer vision and pattern recognition, medical informatics, and human-computer interaction.

Further information can be found in the ImageCLEF book [9] describing the formation, growth, resources, tasks, and achievements of ImageCLEF.

Table 1. Participation in the ImageCLEF tasks and number of participants by year

Task	2003	2004	2005	2006	2007	2008	2009	2010
General images								
Photographic retrieval	4	12	11	12	20	24	19	–
Interactive image retrieval	1	2	2	3	–	6	6	–
Object and concept recognition				4	7	11	19	17
Wikipedia image retrieval						12	8	13
Robot vision							7	7
Medical images								
Medical image retrieval		12	13	12	13	15	17	16
Medical image annotation			12	12	10	6	7	–
Total (distinct)	4	17	24	30	35	45	65	47

3   Bibliometric Analysis Method

Bibliometric studies provide a quantitative and qualitative indication of the scholarly impact of research by examining the number of scholarly publications derived from it and the number of citations these publications receive. The most comprehensive sources for publication and in particular for citation data are: (i) the Thomson Reuters Web of Science[9] (generally known as ISI Web of Science or *ISI*), established by Eugene Garfield in the 1960s, (ii) *Scopus*[10], introduced by Elsevier in 2004, and (iii) the freely available *Google Scholar*[11], developed by Google in 2004. In addition to publication and citation data, ISI and Scopus also provide citation analysis tools to calculate various metrics of scholarly impact, such as the h–index [5], a robust metric of scientific research output that has a value h for a dataset of N_p publications, if h of them have at least h citations each, and the remaining $(N_p - h)$ publications have no more than h citations each. Google Scholar on the other hand is simply a data source and does not have such capabilities; citation analysis using its data can though be performed by the Publish or Perish[12] (PoP) system, a software wrapper for Google Scholar.

Each of these sources follows a different data collection policy that affects both the publications covered and the number of citations found. ISI has a complete coverage of more than 10,000 journals going back to 1900, but its coverage of conference proceedings or other scholarly publications, such as books, is very limited or non–existent. For instance, in the field of computer science, ISI only indexes the conference proceedings of the Springer Lecture Notes in Computer Science and Lecture Notes in Artificial Intelligence series. The citations found are also affected by its collection policy, given that in its General Search, ISI provides only the citations found in ISI-listed publications to ISI–listed publications. Scopus aims to provide a more comprehensive coverage of research literature by indexing nearly 18,000 titles from more than 5,000 publishers, including conference proceedings and 'quality web sources'. In its General Search, it lists citations in Scopus–listed publications to Scopus–listed publications, but only from 1996 onwards. Google Scholar, on the other hand, has a much wider coverage since it includes academic journals and conference proceedings that are not ISI– or Scopus–listed, and also books, white papers, and technical reports, which are sometimes higly cited items as well.

As it is evident, these differences in coverage can enormously affect the assessment of scholarly impact metrics; the degree to which this happens varies among disciplines [1,4]. For computer science, where publications in peer–reviewed conference proceedings are highly valued and cited in their own right, without necessarily being followed by a journal publication, ISI greatly underestimates the number of citations found [10,1], given that its coverage of conference proceedings is only very partial, and thus disadvantages the impact of publications. For example, a recent study examining the effect of using different data sources for citation analysis across different disciplines [4] found that for a particular case of

---

[9] http://apps.isiknowledge.com/   [11] http://scholar.google.com/

[10] http://www.scopus.com/   [12] http://www.harzing.com/pop.htm

an established computer science academic, Scopus found 62% more publications and 43% more citations than ISI. Scopus' broader coverage can though be hindered by its lack of coverage before 1996, but this is not a problem in our case since the ImageCLEF evaluation campaign started in 2003. Google Scholar offers an even wider coverage than Scopus and thus further benefits citation analyses performed for the computer science field [10,4]. As a result, this study employs both Scopus and Google Scholar (in particular its PoP wrapper) for assessing the scholarly impact of ImageCLEF. This allows us to also explore a further goal: to compare and contrast these two data sources in the context of such an analysis. Scopus and Google Scholar were also employed in the examination of the TRECVID scholarly impact [13], where emphasis was though mostly given on the Google Scholar data.

It should be noted that the reliability of Google Scholar as a data source for bibliometric studies is being received with mixed feelings [1], and some outright scepticism [7,8]. This is due to its widely reported shortcomings [10,7,8,1], which mainly stem from its parsing processes. In particular, Google Scholar frequently has several entries for the same publication, e.g., due to misspellings or incorrectly identified years, and therefore may deflate its citation count [10,7]. This though can be rectified through PoP which allows for the manual merging of entries deemed to be equivalent. Inversely, Google Scholar may also inflate the citation count of some publications, since it may group together citations of different papers, e.g., the conference and journal version of a paper with the same or similar title or its pre–print and journal versions [10,7]. Furthermore, Google Scholar is not always able to correctly identify the publication year of an item [7]. These deficiencies have been taken into account into our analysis and addressed with manual data cleaning when possible, but we should acknowledge that the validity of the citations in Google Scholar were not examined, since this is beyond the scope of this study. Next, we describe the collection and analysis of ImageCLEF publications and their citations using Scopus and PoP.

4   The Dataset of ImageCLEF Publications

CLEF's annual evaluation cycle culminates in a workshop where participants of all CLEF tracks (referred to as labs since 2010) present and discuss their findings with other researchers. This event is accompanied by the CLEF workshop proceedings, known as *working notes*, where research groups publish, separately for each track, notebook papers that describe the techniques used in their participation and results. In addition, the organisers of each track (and/or each task within each track) publish overview papers that present the evaluation resources used, summarise the approaches employed by the participating groups, and provide an analysis of the main evaluation results. The papers in the CLEF working notes are available online on the CLEF website and while they are not refereed, the vast majority of participants take the opportunity to publish there.

After the workshop, participants are invited to publish more detailed descriptions of their approaches and more in–depth analyses of the results of their participation, together with further experimentation, if possible, to the *CLEF proceedings*. These papers go through a reviewing process and the accepted ones, together with updated versions of the overview papers, are published in a volume of the Springer Lecture Notes in Computer Science series in the year following the workshop and the CLEF evaluation campaign. That means that the CLEF proceedings of the CLEF 2005 evaluation campaign were published in 2006. This publication scheme was followed until 2009; in 2010 the format of CLEF changed and the participants' and overview papers were only published in the CLEF working notes, i.e., there were no follow–up CLEF proceedings.

Moreover, CLEF participants may extend their work and publish in journals, conferences, and workshops. The same applies for research groups from academia and industry that, while not official participants of the CLEF activities, may decide at a later stage to use CLEF resources to evaluate their approaches. These *CLEF–derived* publications are a good indication of the impact of CLEF beyond the environment of the evaluation campaign. Furthermore, researchers directly involved with the development of CLEF evaluation resources and/or the coordination of tracks and tasks also publish elsewhere detailed descriptions of the applied methodologies, analyses of the reliability of the created resources, and best practices. These *CLEF resources* publications can be seen as complementary to the overview papers in the CLEF proceedings and working notes.

To assess the scholarly impact of ImageCLEF, bibliometric analysis can be applied to the dataset of publications that contains (i) the ImageCLEF–related publications in the *CLEF working notes* and (ii) in the *CLEF proceedings*, (iii) papers describing *ImageCLEF resources* (typically written by ImageCLEF organisers/coordinators), and (iv) *ImageCLEF–derived* publications where ImageCLEF datasets are employed for evaluating the research that is carried out. In this study, the dataset of publications that is analysed is formed as follows:

- *CLEF working notes*: Although publications in the CLEF working notes do attract citations (as discussed in the next section), given that Scopus does not index them, they are excluded from our analysis, so as to allow a "fair" comparison between the two citation data sources.
- *CLEF proceedings*: These publications are indexed by both Scopus and Google Scholar and therefore are included in our analysis. They were located by submitting a separate query for each of the CLEF proceedings published from 2004 to 2010 (and thus corresponding to the CLEF campaigns from 2003 to 2009, respectively). In Scopus, the query "SRCTITLE(lecture notes in computer science) AND VOLUME(*CLEF_proceedings_volume*) AND ALL (image OR photo OR imageclef* OR Flickr)" was entered in the Advanced Search. In PoP, the CLEF proceedings title was used in the Publication field, "image" in the Keywords field, and the publication year of the proceedings in the Year field. In both cases, the results were manually refined and cross–checked against the proceedings, so as to ensure that both the precision and recall of these results are 100%.
- *ImageCLEF resources*: Given that these publications are written by ImageCLEF organisers, they were located by searching by author name. The

results were manually refined by an expert in the field and added to the dataset of publications to be analysed.

- *ImageCLEF-derived publications*: Locating all publications that use Image-CLEF data is a hard task. One may assume that such papers would cite the overview article of the corresponding year of ImageCLEF, but often only the URL of the benchmark is mentioned, or that such papers are written by researchers having access to the data. Both such searches in Scopus and PoP require extensive manual data cleaning and the inclusion of such publications in the analysis is left as part of the next stage of our investigation.

Therefore, this preliminary study to assess the scholarly impact of ImageCLEF focusses on the analysis of the dataset of publications published between 2004 and 2010 and consisting of (i) ImageCLEF–related participants' and overview papers in the CLEF proceedings, and (ii) overview papers regarding ImageCLEF resources published elsewhere. The results are presented in Table 2 and are analysed in the next section.

5   Results for ImageCLEF Publications 2004–2010

The results of our study, presented in Table 2, show that there were a total of 195 ImageCLEF-related papers in the CLEF proceedings published between 2004 and 2010. Over the years, there is a steady increase in such ImageCLEF publications, in line with the continuous increase in participation and in the number of offered tasks (see Table 1). The coverage of publications regarding ImageCLEF resources varies greatly between Scopus and Google Scholar, with the former indexing a subset that contains only 57% of the publications indexed by the latter. These publications peak in 2010, which coincides with the year that ImageCLEF organised a benchmarking activity as a contest in the context of the International Conference for Pattern Recognition (ICPR). This event was accompanied by several overview papers describing and analysing the Image-CLEF resources used in the contest, published in the ICPR 2010 [6] and ICPR 2010 Contests [14] proceedings.

The number of citations varies greatly between Scopus and Google Scholar. For the publications in the CLEF proceedings, Google Scholar finds almost nine times more citations than Scopus. Apart from the wider coverage of Google Scholar, this is also partly due to its inability to distinguish in some cases publications with the same or similar title published in different venues, as is sometimes the case with papers published in the CLEF working notes and in the CLEF proceedings. Differentiating between the citations of two such versions of a CLEF paper requires extensive manual data cleaning that examines the list of references in the citing papers, a task which is beyond the scope of this study. Nevertheless, the inclusion of the citations to the CLEF working notes versions of some CLEF proceedings papers is considered acceptable in the context of this analysis, since they are still indicative of ImageCLEF's scholarly impact. When examining the distribution of citations over the years, Scopus indicates a variation in the number of citations, while Google Scholar shows a relative

Table 2. Overview of ImageCLEF publications 2004–2010 and their citations

	Year	CLEF proceedings			ImageCLEF resources			All		
		papers	citations	h-index	papers	citations	h-index	papers	citations	h-index
Scopus	2004	5	13	2	4	31	3	9	44	4
	2005	20	50	4	–	–	–	20	50	4
	2006	25	24	3	3	28	1	28	52	3
	2007	27	25	2	6	29	2	33	54	3
	2008	29	18	3	5	22	2	34	40	3
	2009	45	14	2	2	4	1	47	18	2
	2010	44	38	4	11	7	2	55	45	4
	Total	195	182	6	31	121	5	226	303	9
Google Scholar	2004	5	65	3	5	105	4	10	170	6
	2005	20	210	8	5	47	4	25	257	10
	2006	25	247	7	8	144	5	33	391	9
	2007	27	259	7	10	76	4	37	335	9
	2008	29	249	7	7	73	5	36	322	9
	2009	45	284	7	7	53	4	52	337	9
	2010	44	259	7	12	76	6	56	335	10
	Total	195	1573	18	54	574	13	249	2147	22

stability from 2005 onwards. For publications regarding ImageCLEF resources, Google Scholar finds almost five times more citations than Scopus. These peak for papers published in 2006 and 2004, mainly due to three publications that describe the creation of test collections that were used extensively in Image-CLEF in the following years, and thus attracted many citations. Overall, Google Scholar indicates that the total number of citations over all 249 publications in the considered dataset are 2,147, resulting in 8.62 average cites per paper. This is comparable to the findings of the study on the scholarly impact of TRECVid [13], with the difference that they consider a much larger dataset of publications that also includes all TREC–derived papers.

Next, we analyse the distribution of citations over different types of papers starting with a comparison of the participants' papers in the CLEF proceedings with overviews describing ImageCLEF resources published both in the CLEF proceedings and elsewhere. Figure 1(a) compares the relative number of papers with the relative citation frequency for these publication types. While participants' papers account for a substantial share of the publications, namely 74.8% for Scopus and 67.9% for Google Scholar, they receive around 35% of the citations. Even when considering only the CLEF proceedings, i.e., when excluding the ImageCLEF resources papers published elsewhere so as to limit the bias towards overview papers that comes from including this dataset in the analysis, Figure 1(b) indicates that while participants' publications constitute 86.7% of the total, they attract around 50% of the citations. These results indicate the significant impact of the ImageCLEF overview papers.

As an example, Figure 2 shows the evolution of the citations for the 2005 overview paper [3]. This is the last paper describing both medical and general tasks in a single overview, and as such it has been cited often. It shows a peak

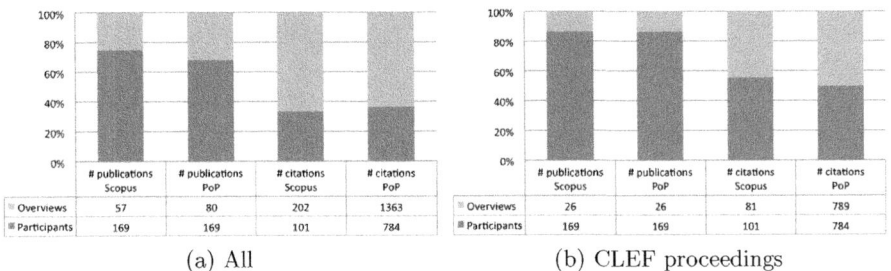

(a) All   (b) CLEF proceedings

Fig. 1. Relative impact of ImageCLEF publication types

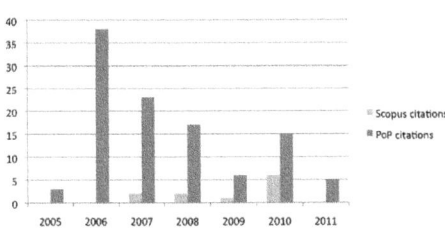

Fig. 2. Citations trends for the 2005 overview paper [3]

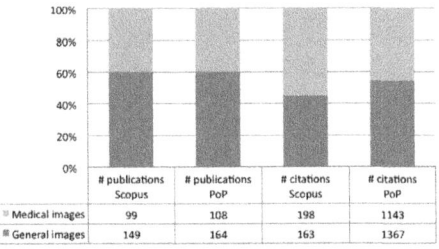

Fig. 3. Relative impact of ImageCLEF publications in the two domains

in the year after the competition and then slowly decreases with a half life of approximately three years in Google Scholar. In Scopus, the peak appears later and in general the number of citations remains almost stable over the years.

Next, the impact of publications in the two domains studied in ImageCLEF, medical and general images, is investigated. Figure 3 compares the relative number of publications with the citation frequency for the domains. It should be noted that some publications examine both domains at once, e.g., participants' papers presenting their approaches in ImageCLEF tasks that represent both domains or overview papers reporting on all tasks in a year. Therefore, the sum of publications (citations) in Figure 3 is not equal to the total listed in Table 2. Overall, the publications in the medical domain appear to have a slightly higher impact. To gain further insights, Figure 4 drills down from the summary data

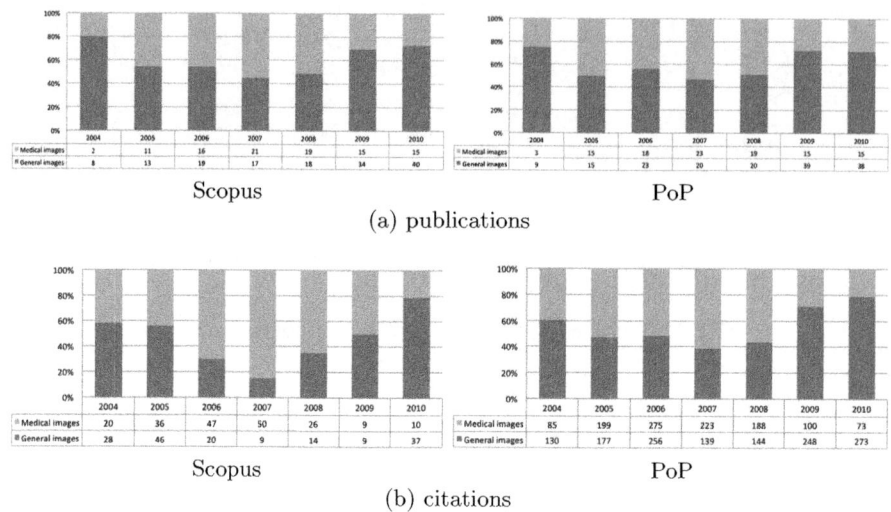

Scopus   PoP

(a) publications

Scopus   PoP

(b) citations

Fig. 4. Relative impact of ImageCLEF publications in the two domains over the years

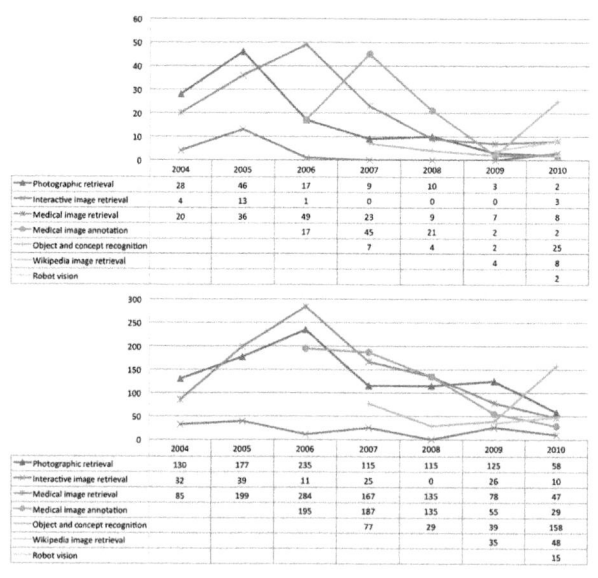

Fig. 5. Citation trends per ImageCLEF task, Scopus (top) and PoP (bottom)

into the time dimension. At first, publications relating to the general domain dominate, with those relating to the medical domain increasing as the corresponding tasks establish themselves in the middle of the time period, while more recently there is again a shift towards the general domain. Scopus indicates that the impact of ImageCLEF publications that are related to the medical domain is particularly significant between 2006 and 2008. This is mostly due to number of overview papers regarding the medical image annotation task published both in the CLEF proceedings and elsewhere, and also because Scopus does not index some of the ImageCLEF publications regarding general images that are found by Google Scholar. For Google Scholar, on the other hand, the distribution of citations appears to be mirroring that of the publications in the two domains.

Finally, Figure 5 depicts the distribution of citations for each of the Image-CLEF tasks (listed in Table 2) over the years. Similarly to above, a publication may cover more than one task. For all tasks, there is a peak in their second or third year of operation, followed by a decline. The exception is the object and concept recognition task, which attracts significant interest in its fourth year when it is renamed as photo annotation task and employs a new collection consisting of Flickr images and new evaluation methodologies. These novel aspects of the task result not only in increased participation (see Table 2), but also strengthen its impact. Overall, the photographic retrieval, the medical image retrieval, and the medical image annotation tasks have had the greatest impact.

6   Conclusions

This paper aims at analysing the scholarly impact of the ImageCLEF image retrieval evaluation campaign. Both Scopus and Google Scholar are used to obtain the number of papers published in the course of ImageCLEF and their citations. This preliminary analysis concentrates on the CLEF post–workshop proceedings, as the CLEF working notes are not indexed by Scopus, and therefore a fair comparison between Scopus and Google Scholar, one of the goals of this study, would not have been possible. A few additional papers written by the organisers about the main workshop outcomes are added. A total of 249 publications were analysed obtaining 2,147 citations in Google Scholar and 303 in Scopus.

A comparison of Google Scholar and Scopus shows that both systems have advantages and limitations. Whereas Scopus is incomplete and misses many conference/workshop papers, the quality of its citation data is high. On the other hand Google Scholar is more complete, but contains errors such as combining publications with similar titles or having two entries for some publications.

An impact analysis over time shows that the half life of citations is around three years for the overview papers. The analysis also shows that tasks usually take a year to attract a larger number of participants but impact and participation usually drop after three years unless the task or the collection changes.

This analysis is only an intermediate step and it seems necessary to extend it to include not only the CLEF proceedings, but also the working notes and derived work. With the proceedings covering almost 230 papers and the non–reviewed working notes a larger number, 500 articles have already been published in this context. Taking into account the derived work, over 1,000 articles can be expected to be based on ImageCLEF data. It is also important to assess the impact of all of CLEF that contains 4–10 tasks and has run for 11 years, already.

This preliminary analysis shows ImageCLEF's significant scholarly impact through the substantial numbers of its publications and their received citations. ImageCLEF data has been used by over 200 research groups, many techniques have been compared during its campaigns, while its influence through imposing a solid evaluation methodology and through use of its resources goes even further.

Acknowledgements. The work was partially supported by the EU in the context of Promise (258191), Chorus+ (249008), and Khresmoi (257528) FP7 projects.

References

1. Bar-Ilan, J.: Which h-index? A comparison of WoS, Scopus and Google Scholar. Scientometrics 74(2), 257–271 (2008)
2. Cleverdon, C.W.: The evaluation of systems used in information retrieval. In: Proceedings of the International Conference on Scientific Information, vol. 1, pp. 687–698. National Academy of Sciences, National Research Council (1959)
3. Clough, P., Müller, H., Deselaers, T., Grubinger, M., Lehmann, T.M., Jensen, J., Hersh, W.R.: The CLEF 2005 cross–language image retrieval track. In: Peters, C., Gey, F.C., Gonzalo, J., Müller, H., Jones, G.J.F., Kluck, M., Magnini, B., de Rijke, M., Giampiccolo, D. (eds.) CLEF 2005. LNCS, vol. 4022, pp. 535–557. Springer, Heidelberg (2006)
4. A.-W. Harzing. Citation analysis across disciplines: The impact of different data sources and citation metrics (2010), http://www.harzing.com/data_metrics_comparison.htm (retrieved)
5. Hirsch, J.E.: An index to quantify an individuals scientific research output. Proceedings of the National Academy of Sciences (PNAS) 102(46), 16569–16572 (2005)
6. Proceedings of the 20th International Conference on Pattern Recognition (ICPR 2010). IEEE Computer Society, Instanbul (2010)
7. Jacsó, P.: Deflated, inflated and phantom citation counts. Online Information Review 30(3), 297–309 (2006)
8. Jacsó, P.: The pros and cons of computing the h-index using Google Scholar. Online Information Review 32(3), 437–452 (2008)
9. Müller, H., Clough, P., Deselaers, T., Caputo, B. (eds.): ImageCLEF: Experimental Evaluation in Visual Information Retrieval, 1st edn. Springer, Heidelberg (2010)
10. Rahm, E., Thor, A.: Citation analysis of database publications. SIGMOD Record 34, 48–53 (2005)
11. Rowe, B.R., Wood, D.W., Link, A.N., Simoni, D.A.: Economic impact assessment of NIST's Text REtrieval Conference (TREC) Program. Technical Report Project Number 0211875, RTI International (2010)
12. Sanderson, M.: Test collection based evaluation of information retrieval systems. Foundations and Trends in Information Retrieval 4, 247–375 (2010)
13. Thornley, C.V., Johnson, A.C., Smeaton, A.F., Lee, H.: The scholarly impact of TRECVid (2003–2009). JASIST 62(4), 613–627 (2011)
14. Ünay, D., Çataltepe, Z., Aksoy, S. (eds.): Proceedings of the 20th International Conference on Recognizing Patterns in Signals, Speech, Images, and Videos, ICPR Contest Reports. Springer, Heidelberg (2010)

The Importance of Visual Context Clues in Multimedia Translation

Christopher G. Harris[1] and Tao Xu[2]

[1] Informatics Program, The University of Iowa, Iowa City, IA 52242 USA
christopher-harris@uiowa.edu
[2] School of Foreign Languages, Tongji University, Shanghai 200092 China
michaeltjxt@hotmail.com

Abstract. As video-sharing websites such as YouTube proliferate, the ability to rapidly translate video clips into multiple languages has become an essential component for enhancing their global reach and impact. Moreover, the ability to provide closed captioning in a variety of languages is paramount to reach a wider variety of viewers. We investigate the importance of visual context clues by comparing transcripts of multimedia clips (which allow transcriptionists to make use of visual context clues in their translations) with their corresponding written transcripts (which do not). Additionally, we contrast translations produced using crowdsourcing workers with those made by professional translators on cost and quality. Finally, we evaluate several genres of multimedia to examine the effects of visual context clues on each and demonstrate the results through heat maps.

1 Introduction

Multimedia content-sharing websites invite global users to discover and share original video content. These websites have become immensely popular as demonstrated by their rapid growth; one such website, YouTube, has quickly grown to be the world's most popular video site since its introduction in early 2005. Each day over a billion video plays are initiated by millions of users on YouTube and similar websites [1]. These video content-sharing websites provide a forum for people to engage with multimedia content globally and also act as an important distribution platform for content creators.

However, their global reach is often limited by restrictions on their ability to translate into other languages by overdubbing or closed captioning. Original content creators are often individuals or small groups without access to substantial capital, limiting the opportunities to have their multimedia creations translated professionally. Also, with so many content creators vying for the limited time and attention of potential viewers, competition is keen; therefore, translations need to be performed quickly so as to capture and enhance international viewer interest before it wanes.

Currently there are several viable methods to translate multimedia content, three of which we explore in this paper. One method is to hire a professional translator to translate into several languages; however, this is costly and thus impractical for most contributors. Another method is to use machine (MT) translational tools, such as

P. Forner et al. (Eds.): CLEF 2011, LNCS 6941, pp. 107–118, 2011.
© Springer-Verlag Berlin Heidelberg 2011

Google Translate, but the quality of these tools is not yet high enough to provide translations for complex concepts in other languages correctly. A third method is to obtain translation through the use of crowdsourcing, which in theory, permits translations to be performed quickly, correctly, and relatively inexpensively. In this paper, we will use the professional translation as our gold standard and explore MT and crowdsourcing approaches.

Evaluating these approaches raises some important issues on the use of visual context cues in multimedia translation. First, is it sufficient to work from a written transcript, as MT tools are required to do, or are the visual context cues found in video truly beneficial for translation quality? Second, since crowdsourcing makes use of humans who can make use of these visual cues, how does crowdsourcing compare to the MT tools available today? Third, how dependent are the previous two questions on the genre of multimedia we choose to translate?

The paper is organized as follows: In the next section, we discuss related work in this area. Section 3 contains a discussion of the Meteor MT evaluation system. In Section 4, we discuss our methodology and experimental approach. In Section 5, we present and discuss our findings and their resulting implications. We conclude and discuss directions for future work in Section 6.

2   Background and Motivation

The use of visual context as an aid to understanding is well covered in the literature. A number of visual contextual studies, (e.g.[2, 3]) have been conducted in the field of computational psycholinguistics, an area of linguistics concerned with the development of computational models that examine how language processing occurs in the brain. In addition, studies involving the utility of visual contextual cues have been explored for their effects on learning and attention [4-7], listening comprehension [8-10], speech and language comprehension [11-13], and second language retention [14-16]; in each of these studies, clear benefits of visual contextual clues have been identified.

Multimedia translations are normally straightforward extensions of the above, and therefore limited research has been focused on the use of visual context; however, some studies have found that translations involving creative imagery can possibly mislead [17] or unintentionally misinform [18]. In addition, if a user unfamiliar with a language observes imagery that is intended to be confused with the regular context of that word (i.e., a "play on words"), it is understandable how visual contextual clues can hinder, not aid, language comprehension as indicated in [19]. Therefore, we examine the use of imagery to examine this aspect.

Most machine translation (MT) tools work by using linguistic rules, using corpus statistics, using examples, or a combination of these techniques to determine the correct inter-lingual substitution between words or phrases. Many of these MT tools are freely available on the internet, such as Google Translate[1], Bing Translator[2] and Babelfish[3]. Although acceptable for simple translation tasks, the quality of MT tools

---

[1] translate.google.com
[2] microsofttranslator.com
[3] www.babelfish.com

is still too poor to use in a professional setting [20]. In fact, since the early 1960s, several doubts have been expressed about the ability to ever achieve fully-automated MT of high-quality [21] and that perfect translations would never be achievable [22].

We also investigate translations using crowdsourcing. Since its introduction in 2006, crowdsourcing has become a viable platform for the "crowd" - a large pool of semi-anonymous users - to perform a set of structured tasks [23]. Crowdsourcing marketplaces, such as Amazon's Mechanical Turk[4] are designed as a labor clearinghouse for "micro-tasks" – small tasks that can be done by anyone who meets preset qualification criteria – in return for payments based on number of tasks completed. Crowdsourcing focuses on micro-tasks requiring human intelligence, and therefore is applicable to the field of translation: work done by crowdsourced workers can be accomplished quickly, inexpensively, and has been demonstrated to be good quality [24], particularly if these micro-tasks are clearly defined and multiple participants perform the same task as a quality check [25].

We examine the use of crowdsourcing in this study since it provides an inexpensive yet reliable substitute for professional translation services [26]. Indeed, a recent study found that crowdsourcing speech transcriptions was nearly as reliable as professional translations but at 1/30th the cost [27]. More importantly, crowdsourcing allows us to examine the role of visual context clues in multimedia translation, which we cannot do with online MT tools.

3   The Meteor MT Evaluation System

We now turn our attention to the evaluation tools for translations. The evaluation tool we use in this study, Meteor [28], was introduced in 2004 to tackle some of the issues other MT evaluation systems did not adequately address. Meteor was designed to improve correlation with human judgments of MT quality at the segment (sentence) level; it has been shown to correlated better with human judgments than other MT systems [28]. Meteor evaluates a translation by computing a score based on explicit word-to-word matches between the translation and a given reference translation. If more than one reference translation is available, the translation is scored against each reference independently, and the best scoring pair is used. Alignments are built incrementally in a series of stages using the following Meteor matchers:

- *Exact*: Words are matched if and only if their surface forms are identical
- *Stem*: Words are stemmed using a stemmer, such as the Porter Snowball Stemmer [29] and matched if and only if the stems are identical.
- *Synonym*: Words are matched if they are both members of a synset (synonym set) according to the WordNet database [30]. This ability to use synonym sets is powerful, since the choice of words used by two human translators may be very similar in meaning but not exact. This use of synsets allows for some flexibility in translation that is reflected in the real world. We made extensive use of this feature in our studies.

---

[4] mturk.com

At each stage, one of the above subroutines locates all possible word matches between the two translations using words not aligned in previous stages. An alignment is then identified as the largest subset of these matches in which every word in each sentence aligns to zero or one words in the other sentence. If multiple such alignments exist, the alignment is chosen that best preserves word order by having the fewest crossing alignment links. At the end of each stage, matched words are marked so that they are not considered in future stages. The resultant Meteor alignment used for scoring is defined as the union of all stage alignments.

The Meteor score for a given pairing is computed based on the number of mapped unigrams found between the two strings, m, the total number of unigrams in the translation, t, and the total number of unigrams in the reference, r. Unigram precision is calculated as $P = m/t$ and unigram recall as $R = m/r$. An F-measure, which is the harmonic mean of precision and recall, is then computed [31]:

$$F_{Mean} = \frac{P \cdot R}{\alpha \cdot P + (1 - \alpha) \cdot R}$$

The value of α determines the tradeoff between precision and recall. The precision, recall and F_{Mean} are all based on single-word matches, but the extent to which the word order matches also needs to be considered. Meteor computes a penalty for a given alignment in the following manner. First, the sequence of matched unigrams between the two strings is divided into the fewest possible number of chunks, maximizing the adjacency of matched unigrams in each string and in identical word order. The number of these chunks, ch, and the number of correct matches, m, is then used to calculate a fragmentation fraction $= ch/m$. To illustrate, a candidate translation that is an exact match with the reference document will result in a single chunk. The penalty is then computed as:

$$Penalty = \gamma \cdot (ch/m)^{\beta}$$

Here, the value of γ determines the maximum penalty $(0 \leq \gamma \leq 1)$. The value of β determines the functional relation between fragmentation calculated and the penalty. In practice, we empirically determine the optimal values for α, β, and γ for each language independently.

Although a number of MT evaluation systems exist and have their merits, we chose Meteor for a number of reasons. First, a number of studies have examined Meteor's correlation with human judgments across a number of scenarios. Additionally, we believe Meteor's use of synonyms provides some flexibility in capturing the essence of a translation better than some of the other metrics. Finally, the Meteor source code is well-maintained and readily available, permitting us to adapt the code, in particular the WordNet-based synonym module, to our specific needs.

4 Experimental Design

We used nine multimedia videos; each was considered challenging to translate due to the amount of figurative language they included. Our goal is to observe the effects of several different features on translation quality. We designed our experiments with four separate features: multimedia genre, translation type, language, and whether or not visual context clues were used in translation.

Meteor values used for α, β, and γ for scoring each of the languages is provided in Table 1. They are optimized based on existing research [32] and from our own preliminary studies.

Table 1. Parameters used with Meteor

	English	Spanish	Russian
α	0.95	0.90	0.85
β	0.50	0.50	0.60
γ	0.45	0.55	0.70

We evaluate two hypotheses: first, the use of visual context clues provides an improvement in the Meteor evaluation score compared with using the written transcripts alone. Second, translations provided by crowdsourcing workers can obtain Meteor assessment scores as high as those from professional translators. If both hypotheses are true, this bolsters our overall claim that visual context clues matter in multimedia translation, and that accurate translations can be made at low cost.

Table 2. Mean Document Size, in words, by Language and Multimedia Genre

	Chinese	English	Russian	Spanish
TS	1,219	1,310	1,164	1,369
AN	684	705	624	743
MV	183	198	149	215

We tested our hypotheses as follows. We took nine short videos, each 5-10 minutes in duration, in Mandarin Chinese from three different genres – three talk shows (TS), three animated comedy skits (AN), and three music videos (MV). We chose multimedia videos with highly-figurative language content to ensure challenging translations. From these, we created written transcripts of each video clip in Simplified Chinese. Next, we hired three professional translators to provide transcripts in three different languages – English (EN), Spanish (ES) and Russian (RU). Figure 1 gives an illustration of the steps taken to break this translation study into several groups for each genre.

We conducted two separate studies – one using the raw multimedia to obtain visual cues while others used the transcripts only – in order to compare the difference in Meteor score. These translations were conducted using crowdsourcing (CS), online machine translation tools (MT) and professional translators (PT) – our gold standard. For the crowdsourcing translations, we hired non-professional translators through several crowdsourcing platforms to provide translations from Chinese into our three target languages. We took steps to ensure the same translator was not used to translate both from the multimedia and from the written transcripts alone for the same language pair, as this could introduce bias.

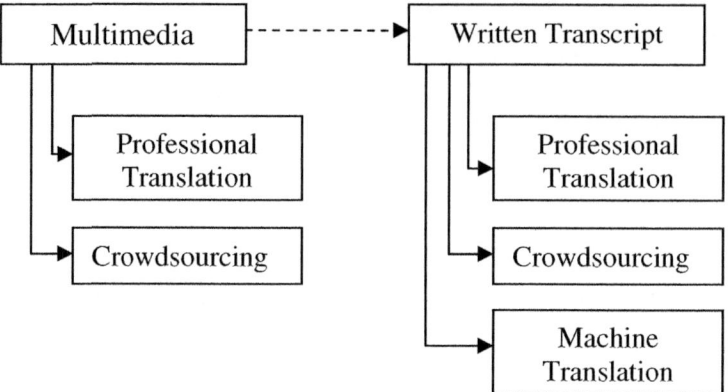

Fig. 1. Overview of the different groups evaluated in this paper

We also use the three aforementioned online MT tools from Google, Babelfish and Bing, as well as two others: Worldlingo[5] and China-based Lingoes[6] to provide translations from the written transcripts into our three target languages. We used the maximum Meteor MT evaluation score obtained from all five online translation tools for a single translation for a single genre. For crowdsourced transcripts, we had a minimum of two translations for each (with an average of 3.8 translations per transcript) and used the maximum score of these in our calculations. We then scored each using Meteor against our (gold standard) professional translation (PT) using the parameters given in Table 1. The version of Meteor we used for our evaluation includes support for English and Spanish. For Russian, we modified the Meteor program a Russian WordNet[7] into our Meteor system for synonym evaluation.

5   Results

Our first hypothesis examined whether the use of visual contextual clues improve the Meteor scores compared with the use of written transcripts alone. When evaluating our results (columns 1-3 and 4-6 in Figure 2), we are unable to conclude that the difference is significant when comparing the group means at $p = 0.05$; however, when we use a paired t-test, we are able to verify a significant difference at the $p = 0.05$ level of confidence.

Our second hypothesis examined the ability for crowdsourced workers to replicate the translational quality of professional translators. We took the professional translator scores as our gold standard, so we could examine the inter-annotator agreement (Cohen's Kappa) between the crowdsourced translations and the professional translations. We consider synonyms for a given term to be equivalent in our scoring. In Table 3, we group our results by multimedia genre instead of language, as this provides a more meaningful examination of their differences.

---

[5] www.worldlingo.com
[6] www.lingoes.cn
[7] www.pgups.ru/WebWN/wordnet.uix

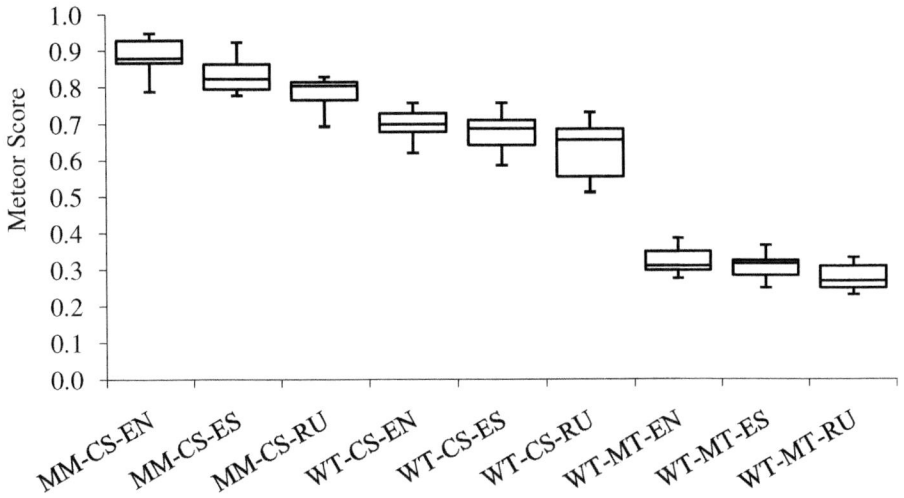

Fig. 2. Overview of Meteor scores comparing the results using all multimedia (MM) and the written transcripts only (WT). These are further grouped by translator type (MT or CS) and language (EN, ES and RU).

We are also able to observe from Figure 2 that there is a discernable difference between Meteor scores from crowdsourced translations (columns 4-6) and those from machine translation tools (columns 7-9), even when we only consider the written transcripts alone. When visual contextual clues are considered (i.e. compare columns 1-3 with columns 7-9 in Figure 2), we notice an even larger contrast, further validating our first hypothesis. The difference in the columns in Table 3 also show a gain in inter-annotator agreement when visual context clues are considered (recall that our gold standard translators had access to the multimedia versions of the video clips and written translations). We notice that the largest improvements are with the Music Videos (MV) – which had the most figurative language and therefore the most difficult for translation from a written document. This ratio of gains was consistent among all three languages.

Table 3. Inter-annotator agreement (Cohen's Kappa) between crowdsourced and professional translations grouped by genre. The MM column considers visual context clues whereas WT only considers the written transcripts.

	MM	**WT**
TS	0.69	0.61
AN	0.71	0.67
MV	0.65	0.57

Our primary interest is to examine differences in our four features: translation type, language, multimedia genre and use of multimedia or written transcripts only;

therefore we also represent our results as heat maps, which convey the differences between many of our features nicely. In Figure 3, we illustrate the difference in scores between the written transcripts for crowdsourced translations and online machine translations in a three-dimensional heat map.

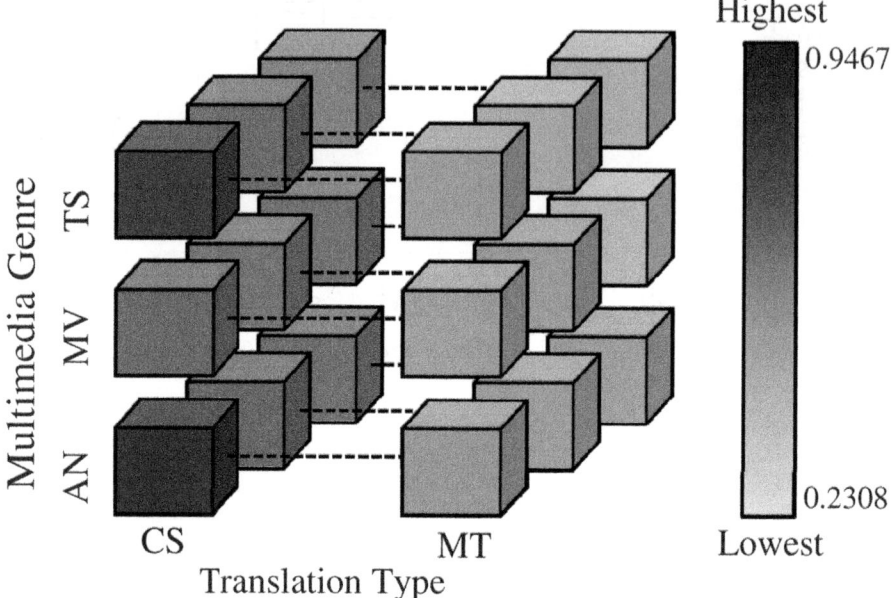

Fig. 3. Meteor scores for translations using visual context clues

From Figure 3, we can observe that the Meteor scores obtained from the crowdsourced translations and machine translations obtained from the written transcripts only are clearly different, and genre and language matter far less than translation type. Figurative language used in these three videos was made clearer through the use of visual context clues in the multimedia content. We also observe that the machine translations have far less variance across genre or language (the colors in the heat map are more uniform) compared with crowdsourced translations.

Figure 4 illustrates the difference in Meteor scores obtained using the visual contextual clues versus using only the written transcripts. The difference in scores ranged from +0.056 to +0.098, demonstrating the benefits of visual context clues as we have discussed earlier. However, with this heat map, we can see that English crowdsourced translations seem to benefit most, particularly those with the animated comedy skits (we believe this is likely due to the use of several situations in those videos involving a "play on words"). The difference in the use of visual context clues was smallest for Russian for the same genre (animated comedy skits). Initially we were not sure if this was due to both the multimedia and the written transcripts accounting for language that involved a play on words, or whether these comedic language constructs were missed entirely in the Russian translations; further examination of the Russian translations indicated the latter.

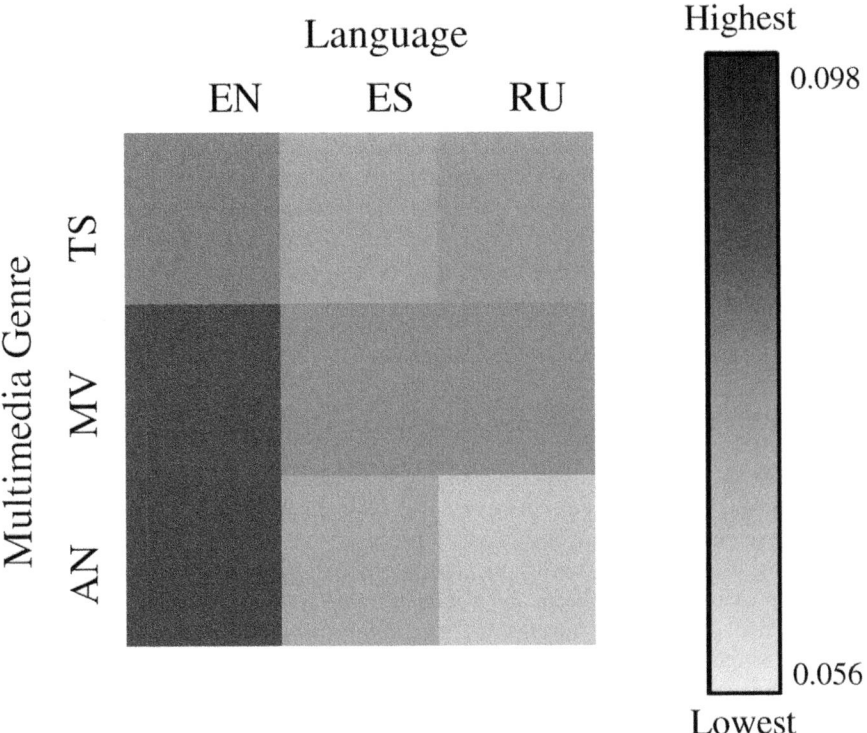

Fig. 4. Meteor scores for translations using visual context clues

Overall, the music videos appear to benefit most from the visual contextual clues. From casual inspection, the three music videos made substantial use of figurative language, which could be clarified through the use of visual aids in the multimedia content. As a result, we believe visual context clues enhance translation ability.

Some differences may be attributable to the Meteor scoring, not to translation ability. For example, "卑微地在人群中隐藏" in one set of lyrics was translated to Russian as "Осмелюсь скрываться в толпе", meaning "I dare to hide in the crowd". In Spanish, this same phrase is directly translated as "Humildemente escondidos en la multitude," the same as its direct translation in English: "Humbly hiding in the crowd." Although the difference in meaning is subtle, the Meteor score difference from the gold standard in the respective language is substantial, pointing out a weakness with Meteor, not with the translation itself. Therefore, to validate translations at a high level, we had human translators provide preference judgments on each feature:

- crowdsourced translations generated from written transcripts compared with crowdsourced translations generated from multimedia
- machine translations compared with crowdsourced translations
- professional translations compared with the crowdsourced translations

These blind preference judgments were able to validate the order of our earlier-reported Meteor rankings on each feature.

One additional issue we wished to validate was the ratio of costs between crowdsourcing translations and professional translations. Our professional translations were done on a per-word basis of 6-12 cents per word, for an average cost of US$49.65 per translation, and took an average of 4.7 business days to complete. Crowdsourcing translations were completed at an average cost of US$2.15 per translation (or 1/23rd of the cost of a professional translation) and took an average of 40 hours (1.6 days) to complete. Even if we require two translations to be done for each document to address any potential quality issues that may arise, we are able to still achieve a substantial savings over professional translations with only a small reduction in quality.

6 Conclusion

Our purpose in this paper was to determine other alternatives to machine translations that are low cost but more effective. We have investigated the role of visual contextual clues in multimedia translations, and the benefits they provide over translations from written transcripts. Additionally, we explored crowdsourcing's ability to provide the fast, cheap, and effective translations described in other studies. We examined these claims through the translation of nine Chinese language videos into three languages. We reported the results through heat maps, which are able to visually represent the relative differences between features.

The results of our study support our two hypotheses. Through a paired t-test, we verified that the visual context clues in our videos were able to increase in the Meteor evaluation scores at a p=0.05 level of significance over translations from written transcripts alone. In addition, we were able to achieve quality translations through crowdsourcing at a fraction of the cost of professional translations, demonstrated by strong inter-annotator agreement scores.

This study represents an initial foray into this area of translation. In the future, we plan to expand our study to include a larger set of videos and more genre variety and examine the role of these translations across a wider variety of languages. This will allow us to determine which languages rely more on visual context clues. In addition, we plan to measure how the translation quality differs between languages from a closely-related family versus languages from more distant families, across different genres.

References

[1] Rao, L.: comScore: YouTube Reaches All-Time High of 14.6 Billion Videos Viewed In (May),   http://techcrunch.com/2010/06/24/comscore-youtube-reaches-all-time-high-of-14-6-billion-videos-viewed-in-may/ (retrieved May 5, 2011)

[2] Crocker, M.: Computational Psycholinguistics. Kluwer Academic Publishing, Dordrecht (1996)

[3]   Grainger, J., Dijkstra, T. (eds.): Visual word recognition: Models and experiments. Computational psycholinguistics: AI and connectionist models of human language processing. Taylor & Francis, London (1996)

[4]   Johnson-Laird, P.N.: Mental Models: Towards a Cognitive Science of Language, Inference, and Consciousness. Cambridge University Press, Cambridge (1983)

[5]   Chun, M.M.: Contextual cueing of visual attention. Trends in Cognitive Sciences 4, 170–178 (2000)

[6]   Torres-Oviedo, G., Bastian, A.J.: Seeing is believing: effects of visual contextual cues on learning and transfer of locomotor adaptation. Neuroscience 30, 17015–17022 (2010)

[7]   Deubel, H., et al. (eds.): Attention, information processing and eye movement control. Reading as a perceptual process. Elsevier, Oxford (2000)

[8]   Mueller, G.: Visual contextual cues and listening comprehension: An experiment. Modern Language Journal 64, 335–340 (1980)

[9]   Meskill, C.: Listening skills development through multimedia. Journal of Educational Multimedia and Hypermedia 5, 179–201 (1996)

[10] Fernald, A., et al. (eds.): Looking while listening: Using eye movements to monitor spoken language comprehension by infants and young children. Developmental Psycholonguistics: On-line methods in children's language processing. John Benjamins, Amsterdam (2008)

[11] Roy, D., Mukherjee, N.: Towards Situated Speech Understanding: Visual Context Priming of Language Models. Computer Speech and Language 19, 227–248 (2005)

[12] Hardison, D.: Visual and auditory input in second-language speech processing. Language Teaching 43, 84–95 (2010)

[13] Cunillera, T., et al.: Speech segmentation is facilitated by visual cues. Quarterly Journal of Experimental Psychology 63, 260–274 (2010)

[14] Long, D.R.: Second language listening comprehension: A schema-theoretic perspective. Modern Language Journal 73 (Spring 1989)

[15] Gullberg, M., et al.: Adult Language Learning After Minimal Exposure to an Unknown Natural Language. Language Learning 60, 5–24 (2010)

[16] Kawahara, J.: Auditory-visual contextual cuing effect. Percept. Psychophys 69, 1399–1408 (2007)

[17] Lew, M.S., et al.: Content-based multimedia information retrieval: State of the art and challenges. ACM Trans. Multimedia Comput. Commun. Appl. 2, 1–19 (2006)

[18] Zhang, X., et al.: A visualized communication system using cross-media semantic association. Presented at the 17th International Conference on Advances in Multimedia Modeling - Volume Part II, Taipei, Taiwan (2011)

[19] Tung, L.L., Quaddus, M.A.: Cultural differences explaining the differences in results in GSS: implications for the next decade. Decis. Support Syst. 33, 177–199 (2002)

[20] Morita, D., Ishida, T.: Collaborative translation by monolinguals with machine translators. Presented at the 14th International Conference on Intelligent User Interfaces, Sanibel Island, Florida, USA (2009)

[21] Bar-Hillel, Y.: A demonstration of the nonfeasibility of fully automatic high quality machine translation. Jerusalem Academic Press, Jerusalem (1964)

[22] Madsen, M.: The Limits of Machine Translation, Masters in Information Technology and Cognition, Scandanavian Studies and Linguistics. University of Copenhagen, Copenhagen (2009)

[23] Howe, J.: The Rise of Crowdsourcing. Wired (June 2006)

[24] Munro, R., et al.: Crowdsourcing and language studies: the new generation of linguistic data. Presented at the NAACL HLT 2010 Workshop on Creating Speech and Language Data with Amazon's Mechanical Turk (CSLDAMT 2010), pp. 122–130 (2010)

[25] Snow, R., et al.: Cheap and fast—but is it good?: evaluating non-expert annotations for natural language tasks. Presented at the Conference on Empirical Methods in Natural Language Processing, Honolulu, Hawaii (2008)

[26] Marge, M., et al.: Using the Amazon Mechanical Turk for transcription of spoken language. In: ICASSP (2010)

[27] Novotney, S., Callison-Burch, C.: Cheap, fast and good enough: automatic speech recognition with non-expert transcription. Presented at Human Language Technologies: The 2010 Annual Conference of the North American Chapter of the Association for Computational Linguistics (HLT 2010), pp. 207–215 (2010)

[28] Banerjee, S., Lavie, A.: METEOR: An Automatic Metric for MT Evaluation with Improved Correlation with Human Judgments. Presented at the ACL Workshop on Intrinsic and Extrinsic Evaluation Measures for Machine Translation and/or Summarization, Ann Arbor, Michigan (2005)

[29] Porter, M.: Snowball: A language for stemming algorithms (2001), http://snowball.tartarus.org/texts/

[30] Miller, G., Fellbaum, C.: WordNet, http://wordnet.princeton.edu (retrieved April 6, 2011)

[31] van Rijsbergen, C.: Information Retrieval, 2nd edn. Butterworths, London (1979)

[32] Agarwal, A., Lavie, A.: METEOR, M-BLEU and M-TER: evaluation metrics for high-correlation with human rankings of machine translation output. Presented at the Third Workshop on Statistical Machine Translation, Columbus, Ohio (2008)

To Re-rank or to Re-query: Can Visual Analytics Solve This Dilemma?

Emanuele Di Buccio[1], Marco Dussin[1], Nicola Ferro[1], Ivano Masiero[1],
Giuseppe Santucci[2], and Giuseppe Tino[2]

[1] University of Padua, Italy
{dibuccio,dussinma,ferro,masieroi}@dei.unipd.it
[2] Sapienza University of Rome, Italy
{santucci,tino}@dis.uniroma1.it

Abstract. Evaluation has a crucial role in Information Retrieval (IR) since it allows for identifying possible points of failure of an IR approach, thus addressing them to improve its effectiveness. Developing tools to support researchers and analysts when analyzing results and investigating strategies to improve IR system performance can help make the analysis easier and more effective. In this paper we discuss a Visual Analytics-based approach to support the analyst when deciding whether or not to investigate re-ranking to improve the system effectiveness measured after a retrieval run. Our approach is based on effectiveness measures that exploit graded relevance judgements and it provides both a principled and intuitive way to support analysis. A prototype is described and exploited to discuss some case studies based on TREC data.

1 Introduction

Inspecting and understanding the causes for the performances of an IR system is always a difficult and demanding task. For example, *failure analysis*, i.e. the detailed and manual analysis for understanding the behaviour and variability of retrieval across topics is often overlooked due to its complexity. The most extensive attempt in this respect has been the Reliable Information Access (RIA) workshop [1] which involved 28 people from 12 organizations for 6 weeks requiring from 11 to 40 person-hours per topic, which shows just how demanding these tasks are.

In this paper, we investigate a methodology for supporting researchers and developers in getting insights about the performances of their algorithms and systems. The methodology builds on the Discounted Cumulative Gain (DCG) family of measures [2,3] because they can handle usefulness scores ranging in a non binary scale and have shown they are especially well-suited both to quantify system performances and to give an idea of the overall user satisfaction with a given ranked list considering the persistence of the user in scanning the list.

We try to better understand what happens when you flip documents with different relevance grades in a ranked list. This is achieved by providing a formal

P. Forner et al. (Eds.): CLEF 2011, LNCS 6941, pp. 119–130, 2011.
© Springer-Verlag Berlin Heidelberg 2011

model that allows us to properly frame the problem and quantify the gain/loss with respect to both an optimal and an ideal ranking, rank by rank, according to the actual result list produced by an IR system. This means that we compare the actual result list with respect to an optimal one created with the same documents retrieved by the IR system, but with an optimal ranking; we also compare the actual result list with respect to an ideal ranking created starting from the relevant documents in the pool (this ideal ranking is what is usually used to normalize the DCG measures). This differs in two ways from what is usually done: firstly, the analysis is conducted rank by rank and not by the overall performances or the area of the difference under two performance curves; secondly, the comparison is done with respect to an optimal ranking created with the same results of the IR system under examination and not only with respect to to an ideal ranking, created with the best results possible, i.e. also considering relevant documents not retrieved by the system.

Our method gives an idea of the distance of an IR system with respect to both its own optimal performances and the best performances possible. The method is adopted as basis for Visual Analytics (VA) techniques that allow analysts to get an intuitive idea through diverse visualizations on possible strategies that could be adopted to improve the IR system performance measured after a retrieval run. In this way, we support researchers and developers in trying to answer an ambitious question: is it better to invest on re-ranking the documents already retrieved by the system or is it better to issue a modified query in the entire collection? In other terms, the proposed techniques allow us to understand whether the system under examination is satisfactory from the recall point of view but unsatisfactory from the precision one, thus possibly benefiting from re-ranking, or if the system also has a too low recall, and thus it would benefit more from re-querying.

Moreover, these visualizations are suitable not only for specialists in the IR field, such as researchers and system developers, but also for users and stakeholders belonging to other communities which employ IR system as components of wider systems. As an example, you can consider the Digital Library (DL) community, where IR systems are usually components of wider DL Systems used to provide access to and retrieval of the multilingual and multimedia cultural heritage assets managed by the system. This is especially important if you consider that such communities which adopt IR system often have difficulties in understanding and assessing the performances of an actual IR system to be embedded into their systems, since this usually requires too specialistic competencies.

This paper describes a prototype that exploits the model and diverse visualizations of it; the prototype is then adopted to analyze several experiments carried out on the TREC7 Ad-hoc track. The paper is organized as follows. Section 2 discusses related work. Section 3 introduces the metrics and the model underling the system together with their visualization and a description of the implemented prototype. Section 4 describes an experiment of the system usage; Section 5 concludes the paper, pointing out ongoing research activities.

2   Related Work

The overall idea of DCG measures is to assign a gain to each relevance grade and for each position in the rank a discount is computed. Then, for each rank, DCG is computed by using the cumulative sum of the discounted gains up to that rank. This gives rise to a whole family of measures, depending on the choice of the gain assigned to each relevance grade and the used discounting function. Typical instantiations of DCG measures make use of positive gains and logarithmic functions to smooth the discount for higher ranks – e.g. a $\log_2$ function is used to model impatient users while a $\log_{10}$ function is used to model very patient users in scanning the result list. More recent works [3] have also tried to assign also negative gains to not relevant documents: this gives rise to performance curves that start falling sooner than the standard ones when non relevant documents are retrieved and let us better grasp, from the user's point of view, the progression of retrieval towards success or failure.

A work that exploits DCG to support analysis is [4] where the authors propose the potential for personalization curve. The potential for personalization is the gap between the optimal ranking for an individual and the optimal ranking for a group. The curves plots the average nDCG's (normalized DCG) for the best individual, group and web ranking against different group size. These curves were adopted to investigate the potential of personalization of implicit content-based and behavior features. Our work shares the idea of using a curve that plots DCG against rank position, as in [2], but using the gap between curves to support analysis as in [4].

The model proposed in this paper provides the basis for the development of VA techniques that can provide us with a quick and intuitive idea of what happened in a result list and what determined its perceived performances. Visual Analytics [5] is an emerging multi-disciplinary area that takes into account both ad-hoc and classical Data Mining (DM) algorithms and Information Visualization (IV) techniques, combining the strengths of human and electronic data processing. Visualisation becomes the medium of a semi-automated analytical process, where human beings and machines cooperate using their respective distinct capabilities for the most effective results. Decisions on which direction analysis should take in order to accomplish a certain task are left to final user. Although IV techniques have been extensively explored [6, 7], combining them with automated data analysis for specific application domains is still a challenging activity [8]. In the VA community previous approaches have been proposed for visualizing and assessing a ranked list of items, e.g. using rankings for presenting the user with the most relevant visualizations [9], or for browsing the ranked results [10].

Visualization strategies have been adopted for analyzing experimental runs, e.g. beadplots in [11]. Each row in a beadplot corresponds to a system and each "bead", which can be gray or coloured, corresponds to a document. The position of the bead across the row indicates the rank position in the result list returned by the system. The same color indicates the same document and therefore the plot makes it easy to identify a group of documents that tend to be ranked near to each other. The colouring scheme uses spectral (ROYGBIV) coding; the ordering adopted for colouring (from dark red for most relevant to light violet for least relevant) is based on a reference system, not on graded judgements and the optimal ranking as in our work. Moreover, in [11] the strategies are adopted for a comparison between the performance of different systems, i.e. the diverse runs; our approach aims at supporting the analysis of a single system, even though it can be generalized for systems comparison.

Another related work is the Query Performance Analyzer (QPA) [12]. This tool provides the user with an intuitive idea of the distribution of relevant documents in the top ranked positions through a *relevance bar*, where rank positions of the relevant documents are highlighted; our VA approach extends the QPA relevance bar by providing an intuitive visualization for quantifying the gain/loss with respect to both an optimal ranking. QPA also allows for the comparison between the Recall-Precision graphs of a query and the most effective query formulations issued by users for the same topic; in contrast, the curves considered in this work allow the comparison between the system performance with the optimal and ideal ranking that can be obtained from a result list.

None of these works deal with the problem of observing the ranked item position, comparing it with an ideal solution, to assess and improve the ranking quality. In [13] the authors explore the basic issues associated with the problem, providing basic metrics and introducing a VA web-based system for exploring the quality of a ranking with respect to an optimal solution. This paper extends such results, allowing for assessing the ranking quality with both the optimal and the ideal solutions and presenting an experiment based on data from runs of the TREC7 Ad-hoc track and the pool obtained in [2].

3   The Formal Model

According to [2] we model the retrieval results as a ranked vector of n documents V, i.e. $V[1]$ contains the identifier of the document predicted by the system to be most relevant, $V[n]$ the least relevant one. The ground truth GT function assigns to each document $V[i]$ a value in the relevance interval $\{0..k\}$, where k represents the highest relevance score. The basic assumption is that the higher the position of a document the less likely it is that the user will examine it, because of the required time and effort and the information coming from the documents already examined. As a consequence, the higher the rank of a relevant document the less useful it is for the user. This is modeled through a discounting function DF that progressively reduces the relevance of a document, $GT(V[i])$ as i increases. We do not stick with a particular proposal of DF and we develop a model that is parametric with respect to this choice. However, to fix the ideas, we recall the original DF proposed in [2]:

$$DF(V[i]) = \begin{cases} GT(V[i]), & \text{if } i \leq x \\ GT(V[i])/\log_x(i), & \text{if } i > x \end{cases} \tag{1}$$

that reduces, in a logarithmic way, the relevance of a document whose rank is greater than the logarithm base. For example, if $x = 2$ a document at position

16 is valuable as one fourth of the original value. The quality of a result can be assessed using the function $DCG(V,i) = \sum_{j=1}^{i} DF(V[j])$ that estimates the information gained by a user who examines the first i documents of V. This paper exploits the variant adopted in `trec_eval` where GT is divided by $\log_x(i+1)$.

The DCG function allows for comparing the performances of different IR systems, e.g. plotting the $DCG(i)$ values of each IR system and comparing the curve behavior. However, if the user's task is to improve the ranking performance of a single IR system, looking at the misplaced documents (i.e. ranked too high or too low with respect to the other documents) the DCG function does not help, because the same value $DCG(i)$ could be generated by different permutations of V and because it does not point out the loss in cumulative gain caused by misplaced elements. To this end, we introduce the following definitions and novel metrics. We denote with $OptPerm(V)$ the set of optimal permutations of V such that $\forall OV \in OptPerm(V)$ it holds that $GT(OV[i]) \geq GT(OV[j])\forall i,j <= n \bigwedge i < j$, that is, OV maximizes the values of $DCG(OV,i)\forall i$. In other words, $OptPerm(V)$ represents the set of the optimal rankings for a given search result.

It is worth noting that each vector in $OptPerm(V)$ is composed of $k+1$ intervals of documents sharing the same GT values. As an example, assuming a result vector composed by 12 elements and $k=3$, a possible sequence of GT values of an optimal vector OV is $< 3,3,3,3,2,2,2,2,1,1,0,0 >$; according to this we define the $max\_index(V,r)$ and $min\_index(V,r)$ functions, with $0 \leq r \leq k$, which return the greatest and the lowest indexes of elements in a vector belonging to $OptPerm(V)$ that share the same GT value r. For example, considering the above 12 GT values, $min\_index(V,2) = 5$ and $max\_index(V,2) = 8$.

Using the above definitions we can define the relative position $R\_Pos(V[i])$ function for each document in V as follows:

$$R\_Pos(V[i]) = \begin{cases} 0, & \text{if } min\_index(V,GT(V[i])) \leq i \leq max\_index(V,GT(V[i])) \\ min\_index(V,GT(V[i])) - i, & \text{if } i < min\_index(V,GT(V[i])) \\ max\_index(V,GT(V[i])) - i, & \text{if } i > max\_index(V,GT(V[i])) \end{cases} \quad (2)$$

$R\_Pos(V[i])$ allows for pointing out misplaced elements and understanding how much they are misplaced: 0 values denote documents that are within the optimal interval, negative values denote elements that are below the optimal interval (pessimistic ranking), and positive values denote elements that are above the optimal (optimistic ranking). The absolute value of $R\_Pos(V[i])$ gives the minimum distance of a misplaced element from its optimal interval.

According to the actual relevance and rank position, the same value of $R\_Pos(V[i])$ can produce different variations of the DCG function. We measure the contributions of misplaced elements with the function $\Delta\_Gain(V,i)$ which compares $\forall i$ the actual values of $DF(V[i])$ with the corresponding values in OV, $DF(OV[i])$: $\Delta\_Gain(V,i) = DF(V[i]) - DF(OV[i])$. Note that while $DCG(V[i]) \leq DCG(OV[i])$ the $\Delta\_Gain(V,i)$ function assumes both positive and negative values. In particular, negative values correspond to elements that are presented too early (with respect to, their relevance) to the user and positive values to elements that are presented too late. Visually inspecting the values

Optimal vector

i	GT(OV)	DF	DCG[i]
1	3	3,00	3,00
2	3	3,00	6,00
3	3	1,89	7,89
4	3	1,50	9,39
5	2	0,86	10,25
6	2	0,77	11,03
7	2	0,71	11,74
8	2	0,67	12,41
9	1	0,32	12,72
10	1	0,30	13,02
11	0	0,00	13,02
12	0	0,00	13,02

Experiment vector

i	GT(V)	DF	DeltaGain	DCG[i]
1	3	3,00	0,00	3,00
2	1	1,00	-2,00	4,00
3	2	1,26	-0,63	5,26
4	3	1,50	0,00	6,76
5	2	0,86	0,00	7,62
6	2	0,77	0,00	8,40
7	3	1,07	0,36	9,47
8	2	0,67	0,00	10,13
9	0	0,00	-0,32	10,13
10	1	0,30	0,00	10,43
11	0	0,00	0,00	10,43
12	3	0,84	0,84	11,27

Fig. 1. Visual representation of R_Pos and Δ_Gain

of these two metrics allows the user to easily locate misplaced elements and understand the impact that such errors have on DCG.

3.1 The Prototype

The results presented in this paper have been implemented in a web based prototype that for a given topic q visualizes the R_Pos and $Delta\_Gain$ functions, together with the DCGs plotted against the rank position for the experiment, the optimal ranking and the ideal ranking where:

Experiment Ranking refers to the top n ranked results provided by the IR approach under consideration;

Optimal Ranking refers to an optimal re-ranking of the experiment ranking where experiment items, namely documents, are ranked in descending order of the degree of relevance according to the judgements in the pool;

Ideal Ranking refers to the top n ranked documents in the pool, where documents are ranked in descending order of their degree of relevance.

Basically, *optimal* refers to the best ranking the system could have provided on the basis of the retrieved documents, while *ideal* refers to the best ranking the system could have provided on the basis of the knowledge of all the relevant documents in the pool. From now on the curves obtained by interpolating the DCGs at the diverse rank positions for the experiment, the optimal, and the ideal ranking will be named respectively *experiment*, *optimal*, and *ideal curve*.

Figure 1 shows the visualization choices adopted in the VA prototype. The leftmost table in the figure represents one of the optimal vectors of $OptPerm(V)$. The second column of the table contains the GT values, the third one the DF values (computed using a log_2 function), and the fourth one the DCG function. The rightmost table represents the experiment result V. The second column

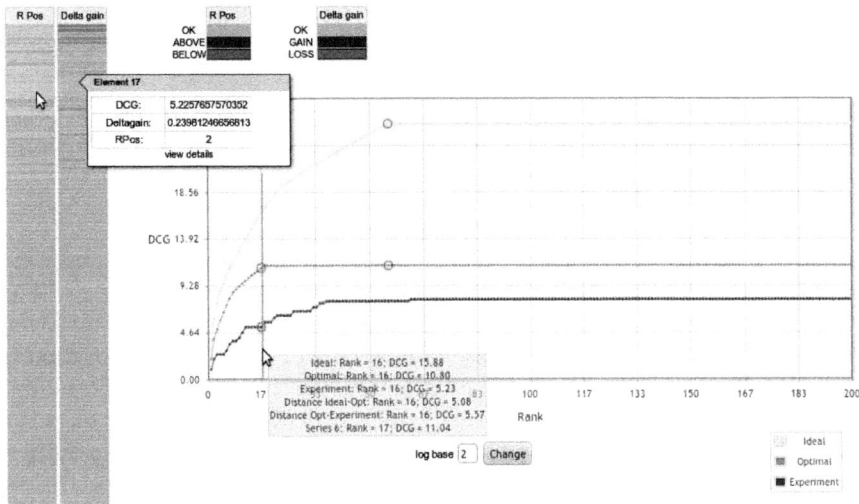

Fig. 2. A screenshot of the prototype

contains the *GT* values together with the *R_Pos* function, coded through color shading: values on correct position=green, values on above positions=blue, and values on below positions=red. The third column contains the *DF* values. The fourth column contains the *Δ_Gain* function, where negative values are coded in red, positive values are coded in blue, and 0 values are coded in green. The fifth column represents the experiment DCG function.

The prototype allows researchers and analysts to compare the experiment result with both the optimal and the ideal result. This facilitates the activities of failure analysis, making it easy to locate misplaced elements, blue or red items, that pop up from the visualization as well as the extent of their displacement and the impact they have on DCG. In this way the analyst can gain insights into the worst errors of the IR system and devise suitable recovering actions.

Figure 2 shows a screenshot of the prototype: the vector on the left represents the *R_Pos* function through color shadings: green, light red/red, and light blue/blue. It allows for locating misplaced documents and, thanks to the shading, understanding how far they are from the optimal position. The vector on the right shows *Delta_Gain* values: light blue/blue codes positive values, light red/red negative values, and green 0 values. A mouse-over triggered interactive pop-up window allows for inspecting the numerical values of single documents: *R_Pos*, *Delta_Gain* and DCG, together with a link to the document. The rightmost part of the screen shows the DCG graphs of the ideal vector, of the optimal vector and of the experiment vector, namely the ranking curves. The points of maximum distance between the experiment and the optimal curves and between the optimal and the ideal curves (highlighted by red circles) can also be seen. A useful popup appears when the mouse is over the graph and displays information about the DCGs of the curves and the distance between them at the rank identified by the mouse position. Brushing allows for highlighting relationships between graph and vectors; indeed, by placing the mouse cursor over colored rows the corresponding point on the graph is highlighted. Finally, through the input panel below the graphs the logarithm base can be changed to model different discount functions according to different classes of search users.

4   Experimentation

The objective of the VA approach introduced in this paper is to support a researcher or analyst investigating how to improve the effectiveness of an IR approach, when the results for one or more queries on the same topic are available. Let us consider, for instance, the case of a retrieval run on a test collection for which the IR approach under evaluation is not effective for one or more topics when considering the top n — in this work $n = 1000$; let us focus on one of these topics. Possible causes of poor performance can be a lack of capability of the system in either: (i) retrieving relevant documents, e.g. a low recall is observed; or (ii) ranking highly relevant document at high rank positions, when the measure of effectiveness adopted is based on graded relevance judgements, as for the family of measures considered in this work; or both of these. A possible approach to address the former issue is to perform a new modified query on the entire collection in order to gather additional relevant documents among the top n. In contrast, if a high recall is observed but the system was unable to rank the documents in descending degree of relevance, a more effective choice to improve the effectiveness of that run could be performing top n document re-ranking in order to achieve the optimal ranking. In this paper we will focus on visualization to support the selection of the strategies to improve system performance, not on the actual implementation of these strategies. In this section we will show how the proposed VA approach can help address the following question: *given a ranked result list obtained in response to a query submitted to the system, should we re-rank the top n documents in the retrieved result list or issue a new modified query on the entire collection?*

The remainder of this section will discuss how the prototype can be adopted to address this question, specifically considering some case studies based on data of the Ad-hoc Track of the TREC7 evaluation campaign.

Dataset. The test collection adopted is based on data from the TREC7 Ad-hoc test collection. A subset of all the *topics* 351-400 is considered, specifically those re-assessed in [2]. Indeed, the *relevance judgements* adopted are those obtained by the evaluation activity carried out in that paper. All the relevant documents of 20 TREC7 topics and 18 TREC8 topics were re-assessed together with 5% of documents judged as not relevant, where assessment was performed using a four graded relevance scale; details on the re-assessment procedure can be found in [2]. The TREC7 Ad-hoc test collection together with this set of judgements were used because of the family of measures adopted in our VA approach, namely DCG. The way the VA approach can be adopted to support researchers and analysts during the evaluation is based on runs submitted to the TREC7 Ad-hoc

Track. In order to be consistent with the choice adopted in [2] we will visualize the curves for top $k = 200$ rank positions.

DCG Curves to Support Per-Topic Analysis. A possible approach for addressing the above research question is to examine the DCG curves, specifically looking at the distance between them. Let us consider, for instance, the experiment KD71010q whose data is visualized in Figure 2 for topic 365. The existence of a gap between the experiment and the optimal curve suggests that an improvement in terms of DCG can be obtained by investigating an optimal re-ranking for the set of retrieved documents for the IR approach under evaluation and the considered topic. Indeed, the distance between experiment and optimal curve at a given rank position indicates the maximum increment in terms of gain that can be achieved by an optimal re-ranking; for instance, at rank 16 the maximum increment that can be achieved in terms of gain is $\Delta = 10.80 - 5.23 = 5.57$ — this Δ differs from Δ_Gain.

In general, if a gap exists between experiment and optimal curve, an improvement in terms of effectiveness can be accomplished by investigating a strategy for optimal re-ranking of the retrieved document set. However re-ranking is not necessarily the best strategy to adopt. Indeed, an analysis of the optimal and the ideal curves reported in Figure 2 shows that a large gap exists between them, which indicates that the system retrieved a low number of relevant documents among those present in the pool, namely a low recall. Therefore, the researcher can opt for investigating strategies based on automatically modified queries, for instance exploiting feedback strategies, and issued on the entire collection, in order to increase the number of relevant documents retrieved instead of trying to optimally re-rank those currently retrieved.

In contrast the curves concerning experiment mds98td and topic 387, whose data are visualized in Figure 3, suggest that investigating the re-ranking strategy can be beneficial if we are interested in improving effectiveness at high rank

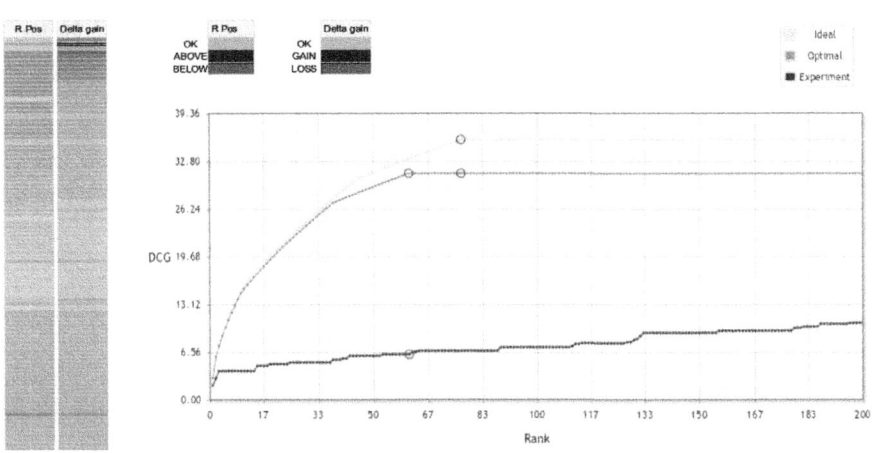

Fig. 3. Curves for experiment mds98td when considering topic 387

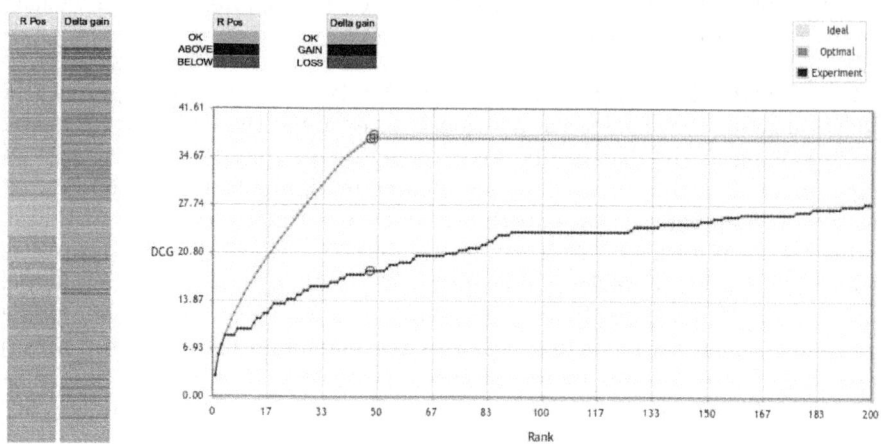

Fig. 4. Curves for experiment Brkly25 when considering topic 358

positions. A gap exists between the optimal and ideal curves, thus indicating that we can further improve recall, but the curves basically overlap in the top 10 rank positions and $\Delta < 1$ is observed for rank positions from 10 to 40. These values suggest that the IR system was able to retrieve highly relevant documents among the top 1000, but it was not able to rank them at high rank positions. The IR system was therefore effective in supporting a first stage prediction, i.e. the first of a series of search episodes needed by the search user in order to achieve his goal in multiple steps. Re-ranking is the best choice in this case. Another example is that depicted in Figure 4. The visualized data concern run Brkly25 and topic 358. Also in this case, the gap between the curves suggests re-ranking could be the best choice: the ideal and optimal curves are overlapped up to rank 40, namely $\Delta_{rank\leq 40} = 0$, and $0.19 < \Delta_{rank>40} < 0.54$. Both runs depicted in Figure 3 and Figure 4 can benefit from re-ranking, as visually suggested by the gap among curves. But the comparison of Δ_Gain vectors, specifically the difference in terms of shading and in number of green entries, shows that there are more documents in the top-most positions that are far from their optimal position in the former case than in the latter. The analyst can interact with the worst ranked documents by a click, thus inspecting the document in order to understand which of its properties were possible causes of failures.

Towards a VA-based Methodology to Support IR Experiment Analysis. The previous section discussed how the described prototype can support an analysis on a per-topic basis. An additional issue is the automatic categorization of topics according to the possible causes of failures of the system when searching from them. The approach we adopted to identify possible topics that can benefit for "re-ranking" or "re-query" is based on the correlation between vectors describing experiment, optimal and ideal ranking. Each topic is described by a pair $(\tau_{ideal-opt}, \tau_{opt-exp})$, where $\tau_{ideal-opt}$ denotes the

Kendall τ correlation between the ideal and the optimal vectors of gains, while $\tau_{opt-exp}$ denotes the Kendall τ between the optimal and experiment vectors of gains. When the pair is $(1,1)$ the best performance is achieved: this is the case of run KD71010q for topic 385. A pair where $\tau_{ideal-opt}$ is high and $\tau_{opt-exp}$ is low suggests that re-ranking could probably improve effectiveness, since there is a strong correlation between ideal and optimal ranking, thus suggesting that the IR approach was quite effective in retrieving relevant documents, but not in the document ranking. This is for instance the case of run mds98td and topic 387 depicted in Figure 3 where the τ pair is $(0.88, 0.07)$. A pair where $\tau_{ideal-opt}$ is low or negative suggests "re-query" on the entire collection as a possible strategy to improve retrieval effectiveness, since an optimal re-ranking of the retrieved document is far from the ideal ranking. This is for instance the case of run KD71010q and topic 365 depicted in Figure 2 where the τ pair is $(0.59, 0.45)$. τ pairs can be adopted in a three step methodology: (i) the pair values allow a first approximation to be obtained when identifying possible causes of failures and topics for which the approach failed; (ii) ranking curves analysis allows for a more in-depth investigation on a per-topic basis, and (iii) Δ_Gain and R_Pos vectors allows for an analysis on a per document basis.

5   Conclusion and Future Work

This paper presents some preliminary results of a VA system for IR evaluation able to explore the quality of a ranked list of documents. The challenging goal of the system is to point out the location and the magnitude of ranking errors in a way that provides insights that contribute to improving the ranking algorithm effectiveness. The system builds on existing and novel metrics that capture the quality of a ranking and allow us to compare it to the optimal one constructed starting from the actual results produced by the system, modeling the degree of satisfaction of a user when s/he inspects those search results. The comparison of the ranking curves, as well as the Δ_Gain and R_Pos vectors, provides an intuitive tool to support IR researchers when conducting retrospective analysis.

Future versions of the prototype could exploit Δ_Gain and R_Pos vector visualization as entry points for more complex user interaction, e.g. manual ranking modification. In [14] we reported on the design and the implementation of a prototype that accesses experimental data via standard Web services from a dedicated system. Access via a web service is adopted in order to allow for the design and development of various client applications and tools for exploiting those data; the prototype described in this paper is an instance of such applications. The prototype in [14] has been developed for a touch device and will be adopted to support the user study we intend to carry out to assess the methodology proposed in this paper. We are currently investigating the metrics, algorithms and visualizations able to locate and visualize the most productive permutations of the result vectors, i.e. heuristic based best flips, and ways of visually correlating the rank of the documents with the ranking algorithm parameters. Lastly, the limitations observed for the Kendall τ in IR evaluation suggest using more suitable variants, e.g. those able to exploit graded relevance scale as proposed in [15].

Acknowledgements. The work reported in this paper has been supported by the PROMISE network of excellence (contract n. 258191) project as a part of the 7th Framework Program of the European commission (FP7/2007-2013).

References

1. Harman, D., Buckley, C.: Overview of the Reliable Information Access Workshop. Information Retrieval 12, 615–641 (2009)
2. Järvelin, K., Kekäläinen, J.: Cumulated gain-based evaluation of IR techniques. ACM Transactions on Information System 20, 422–446 (2002)
3. Keskustalo, H., Järvelin, K., Pirkola, A., Kekäläinen, J.: Intuition-supporting visualization of user's performance based on explicit negative higher-order relevance. In: Proceedings of SIGIR 2008, pp. 675–682. ACM, New York (2008)
4. Teevan, J., Dumais, S.T., Horvitz, E.: Potential for personalization. ACM Transactions on Computer-Human Interaction (TOCHI) 17, 1–31 (2010)
5. Keim, D., Andrienko, G., Fekete, J.D., Görg, C., Kohlhammer, J., Melançon, G.: Information visualization, pp. 154–175. Springer, Heidelberg (2008)
6. Card, S.K., Mackinlay, J.: The structure of the information visualization design space. In: Proceedings of InfoVis 1997, pp. 92–99. IEEE Computer Society, Washington, DC, USA (1997)
7. Shneiderman, B.: The eyes have it: A task by data type taxonomy for information visualizations. In: Proceedings of the 1996 IEEE Symposium on Visual Languages, pp. 336–343. IEEE Computer Society, Washington, DC, USA (1996)
8. Keim, D., Kohlhammer, J., Santucci, G., Mansmann, F., Wanner, F., Schäfer, M.: Visual Analytics Challenges. In: Proceedings of the eChallenges 2009 (2009)
9. Seo, J., Shneiderman, B.: A rank-by-feature framework for interactive exploration of multidimensional data. Information Visualization 4, 96–113 (2005)
10. Derthick, M., Christel, M.G., Hauptmann, A.G., Wactlar, H.D.: Constant density displays using diversity sampling. In: Proceedings of InfoVis 2003, pp. 137–144. IEEE Computer Society, Washington, DC, USA (2003)
11. Banks, D., Over, P., Zhang, N.-F.: Blind Men and Elephants: Six Approaches to TREC data. Information Retrieval 1, 7–34 (1999)
12. Sormunen, E., Hokkanen, S., Kangaslampi, P., Pyy, P., Sepponen, B.: Query performance analyser -: a web-based tool for ir research and instruction. In: Proceedings of SIGIR 2002, p. 450. ACM, New York (2002)
13. Ferro, N., Sabetta, A., Santucci, G., Tino, G., Veltri, F.: Visual comparison of Ranked Result Cumulated Gains. In: Proceedings of EuroVA 2011 (2011)
14. Di Buccio, E., Dussin, M., Ferro, N., Masiero, I., Santucci, G., Tino, G.: Interactive analysis and exploration of experimental evaluation results. In: Proceedings of EuroHCIR 2011 (to appear, 2011)
15. Melucci, M.: Weighted rank correlation in information retrieval evaluation. In: Lee, G.G., Song, D., Lin, C.-Y., Aizawa, A., Kuriyama, K., Yoshioka, M., Sakai, T. (eds.) AIRS 2009. LNCS, vol. 5839, pp. 75–86. Springer, Heidelberg (2009)

Evaluation Methods for Rankings of Facetvalues for Faceted Search

Anne Schuth and Maarten Marx

ISLA, University of Amsterdam, The Netherlands
{anneschuth,maartenmarx}@uva.nl

Abstract. We introduce two metrics aimed at evaluating systems that select facetvalues for a faceted search interface. Facetvalues are the values of meta-data fields in semi-structured data and are commonly used to refine queries. It is often the case that there are more facetvalues than can be displayed to a user and thus a selection has to be made. Our metrics evaluate these selections based on binary relevant assessments for the documents in a collection. Both our metrics are based on Normalized Discounted Cumulated Gain, an often used Information Retrieval metric.

1 Introduction

Search interfaces to semi-structured data often provide ways of refining a full-text search query by selecting values of meta-data fields. These fields —called *facets*— and their values —called *facetvalues*— are then used to filter the results.

We can usually only present a limited number of such facetvalues to a user; both because we have limited amount of space (on a screen) but also because we do not want to put a burden of sifting through a large amount of facetvalues on a user. So, out of all facetvalues a selection has to be made; this paper investigates ways of evaluating such a selection.

In broad terms, we aim at finding a metric that prefers facetvalues that would minimize navigation for a user; a metric that prefers the shortest navigational path through the collection of documents. We want to guide a user in as little as possible steps to all documents that are relevant to his query.

We view the setting in which the selection of facetvalues occurs as follows. We have a collection of documents, some queries and binary relevance judgments (by human assessors) for some documents in the collection for each query, we assume all other documents irrelevant. Besides, we assume that a query defines a strict linear order over the documents.[1] This ordering we assume given. So, for each query we know which documents are relevant and how all documents should be ordered. Also, all our documents are semi-structured, meaning that they contain textual data (on which the ordering is based) as well as meta-data. This meta-data determines to which facetvalues a document belongs.

---

[1] Such an ordering can be based on some similarity score between textual data of the document and the query.

P. Forner et al. (Eds.): CLEF 2011, LNCS 6941, pp. 131–136, 2011.

2   Motivation

While there has been work on evaluation faceted search systems from a user interface perspective (Kules et al., 2009; Burke et al., 1996; English et al., 2002; Hearst, 2006, 2008, 2009), no work has focused on ranking facetvalues from an Information Retrieval perspective. We view our work as enabling research in that direction and would propose doing so in an evaluation campaign with a setup as described below.

Task. Each participant receives the following: a) a collection of queries Q; b) a collection of semi-structured documents D; c) a strict linear order defined on these documents for each query $q \in Q$; and d) a set of facets that may be used, the corresponding facetvalues are dictated by the structured part of the document collection. The task is then to return an ordered list or —depending on the measure— tree of facetvalues that maximizes one of the two metrics defined in this report.

Evaluation. Both our metrics, as described in Section 3, can use simple binary relevance judgments on document per query level. And both return a single number that measures how good a list or tree of facetvalues is. This can be averaged over all queries. To evaluate a participant, the following is needed: a) a collection of semi-structured documents D; b) a strict linear order defined on these documents for each query $q \in Q$; c) binary relevance judgments for each document $d \in D$, for each query $q \in Q$; and d) the submission of the participant: a list or tree of facetvalues for each query $q \in Q$. [2]

3   Two Evaluation Metrics

We introduce two new evaluation metrics. First the Normalized Discounted Cumulated Gain which is an adaption of an existing metric NDCG as described by Järvelin and Kekäläinen (2002). Second, we introduce a new metric called NRDCG which is recursive version of NDCG. Each of our metrics is meant to measure the quality of an *ordered* list or tree of facetvalues. In Table 1 we first introduce some notation, partly inspired by Dash et al. (2008).

3.1   Normalized Discounted Cumulated Gain

This metric focuses on the following two rather loosely formulated aspects: a) prefer facetvalues that would return a lot of relevant documents high in the result list; and b) prefer facetvalues that would return relevant documents we have not seen yet by earlier facetvalues. The first aspect can be measured by counting the number of relevant documents end up in the top p of results if the document-list

---

[2] Note that these requirements imply that at least the INEX 2010 Data Centric Track (Trotman and Wang, 2010) data and relevance judgments can be used with almost no adjustments.

Table 1. Notation used for the definition of our metrics, inspired by Dash et al. (2008)

d	a document, consisting of triple $< t, FV, R >$. With free-text t, set of facetvalues FV and a set of relevance judgments R consisting of $r_q \in \{0, 1\}$ for each query $q \in Q$, where $r_q = 1$ if the document is judged relevant to query q by human judges.
D	list of documents in arbitrary order
D_q	list of documents D ordered by query q
D_f	list of documents D filtered by facetvalue f (in arbitrary order). Or $D_f = \{d : d \in D \wedge f \in FV(d)\}$
f	a facet value pair *facet:value*. A facetvalue is a property of a document, in filtering operations it preserves only those documents that have this property.
F	list of facetvalues.
FT	tree of facetvalues.
q	a full-text query that can define an ordering on D
Q	a set of full-text queries
$D_q[i \dots j]$	list of documents i up to j in the ordered list of documents D_q. Note that the result of $D[\cdot]$ and $D_f[\cdot]$ is arbitrary since the order of those lists is arbitrary.
$t(d)$	the free-text t of document d
$FV(d)$	the set of facetvalues FV of document d
$r(d, q)$	the binary relevance judgment r of document d with respect to query q
$R(D, q)$	list of the relevant documents given a query q that occur in list of documents D. Or $R(D, q) = \{d : d \in D \wedge r(d, q) = 1\}$
n	maximum number of facetvalues per navigation step
p	maximum number of resulting documents in which to look for relevant documents
λ	used in NRDCG to balance direct Gain with Gain in drill-down, $\lambda = 0$ causes NRDCG to reduce to NDCG. $0 \le \lambda < 1$.

D_q were filtered by a facetvalue. The second aspect can then be satisfied by discarding all relevant documents that were already covered by earlier facetvalues, in a ranked list (or tree) of facetvalues. Our notion of Gain, as explained below, is designed to capture both aspects. Because we use a discounted measure we value the Gain of a facetvalue more if it is returned earlier in a ranked list of facetvalues. If we allow the (binary) relevance judgments for documents per query to transfer to facetvalues, we get *graded* relevance for facetvalues. We want these relevance judgments —these gains— to reflect the number of relevant documents in the top p result if a facetvalue were selected. Then, to judge the quality of a ranked list of facetvalues, we could use the Normalized Discounted Cumulated Gain (NDCG) measure (Järvelin and Kekäläinen, 2002) that is designed to evaluate a ranking using graded relevance.

In order to be able to compare DCG for several queries, a normalization step is needed. We can use the regular version of Normalized Discounted Cumulated Gain:

$$NDCG(D, F, q) = \frac{DCG(D, F, q)}{IDCG(D, q)} \tag{1}$$

With a regular definition of Discounted Cumulated Gain:

$$DCG(D, F, q) = \sum_{i=1}^{\min(n, |F|)} \frac{G(D, F, i, q)}{\log_2(i + 1)} \tag{2}$$

So far, nothing was new. Only our definition of Gain is adjusted to reflect the transfer of relevant judgments of documents to facetvalues. We define Gain, $G(D, F, i, q)$ as follows:

$$G(D, F, i, q) = \left| R(D_{q, f_i}[1 \dots p], q) \setminus \bigcup_{j=1}^{i-1} R(D_{q, f_j}[1 \dots p], q) \right| \tag{3}$$

Note how this version of Gain does not take relevant documents covered by earlier facetvalues into account; nothing is gained by returning the same relevant result more than once. It forces the overall measure to prefer facetvalues that cover *new* relevant documents.

Evidently, for the normalization step, we need to calculate how well an ideal ranked list of facetvalues for this query would do; we calculate the Ideal Discounted Cumulated Gain as follows:

$$IDCG(D,q) = \sum_{i=1}^{n} \frac{IG(D,i,q)}{\log_2(i+1)} \tag{4}$$

Where our version of the Ideal Gain states that each ith facetvalue could cover at most p new relevant document:

$$IG(D,i,q) = \max(0, \min(p, |R(D,q)| - (i-1) \cdot p)) \tag{5}$$

Since we normalized the measure for each query, averaging is simple:

$$\overline{NDCG}(D,FT,Q) = \frac{1}{|Q|} \cdot \sum_{q \in Q} NDCG(D,FT,q) \tag{6}$$

3.2   Normalized Recursive Discounted Cumulated Gain

We could also look at the problem of finding the right facetvalues differently. In a real (and possibly even optimal) system a user might have to navigate through multiple facetvalues before he arrives at the desired (relevant) document. In other words, that means that it might take a couple steps before all *relevant* documents end up in the top p results. If we are trying to optimize this navigation we might want to take into account the consecutive facetvalues —and their quality— that a user encounters in a navigation session.

So, instead of looking for the optimal ranked *sequence* of facetvalues, we are looking for an optimal ranked *tree* of facetvalues. We change the setting described in the introduction; we now look for a facetvaluetree FT that optimizes our recursive metric. Such a facetvaluetree FT essentially consists of nested lists of facetvalues and looks like this:

$$FT = (f_1(FT_1), \ldots, f_n(FT_n)) \tag{7}$$

For a tree FT we denote the children of the root with f_i, and we use the notation FT_i to denote the subtree rooted at f_i. The only (natural) restriction on this tree is that paths may not contain a facetvalue more than once.[3]

We define a metric to evaluate this tree in a fashion similar to NDCG, but defined recursively and thus called: Normalized Recursive Discounted Cumulated Gain.

$$NRDCG(D,FT,q) = \frac{RDCG(D,FT,q)}{IRDCG(D,q)} \tag{8}$$

---

[3] This restriction is needed because we only filter out the top p results in the recursive call to $RDCG$, and not *all* documents that are covered by a facetvalue.

We use a recursive version of DCG called Recursive Discounted Cumulated Gain, that is not different except for that it sums up a Recursive Gain.

$$RDCG(D, FT, q) = \sum_{i=1}^{\min(n,|FT|)} \frac{RG(D, FT, i, q)}{\log_2(i+1)} \qquad (9)$$

The Recursive Gain, RG, is a mixture model that is composed of a direct gain borrowed from the normal $NDCG$ and a recursive step. Note that for the recursive call we shrink the document set with those that were displayed already; this causes the measure to focus on unseen (relevant) documents in the remaining steps of a drill-down session.

$$RG(D, FT, i, q) = (1 - \lambda) \cdot G(D, FT, i, q) + \lambda \cdot RDCG(D \setminus D_{q,f_i}[1 \ldots p], FT_i, q) \qquad (10)$$

Setting $\lambda = 0$ reduces the measure to NDCG. Setting $\lambda = 1$ would lead to a zero score (and even division by 0) thus this is not allowed. If $\lambda > 0.5$, the recursive part would get more weight, thereby preferring relevant documents to appear later in a drill-down session. Given that we are after a quick navigation session, we suggest setting $0 < \lambda < 0.5$.

Note that no explicit stopping criteria is needed as $G(\cdot)$ returns 0 for empty document lists and $RDCG(\cdot)$ returns 0 for empty facetvalue lists.

Also, verify that we can indeed simply use $G(D, FT, i, q)$ —even though that function is defined on a list—, as the Gain function is only looking at the children f_i and f_j of the root of FT. Documents (relevant or not) covered by ancestor facetvalues are filtered out by the recursive call to $RDCG(\cdot)$, that is done using $D \setminus D_{q,f_i}[1 \ldots p]$ instead of simply D.

To normalize the $RDCG$, we will need an ideal version, Ideal Recursive Discounted Cumulated Gain ($IRDCG$), which naturally is defined recursively:

$$IRDCG(D, q) = \sum_{i=1}^{\min(n,|R(D,q)|)} \frac{IRG(D, i, q)}{\log_2(i+1)} \qquad (11)$$

$$IRG(D, i, q) = (1 - \lambda) \cdot IG(D, i, q) + \lambda \cdot IRDCG(D \setminus R(D, q)[1 \ldots p], q) \qquad (12)$$

As with $NDCG$, we average over all queries to arrive at $\overline{NRDCG}(D, FT, Q)$.

4   Conclusions

We have have introduced two related measures for evaluating rankings of facetvalues. One might prefer NDCG over NRDCG for its simplicity while the recursive variant might be preferred because it is much more fine grained. Meaning that NRDCG is better at judging a selection of facetvalues were the number of relevant documents is small and harder to retrieve.

Even though we did not include experimental results, our experiments have shown that both measures rank systems that are expected to perform better higher. That is a necessary —not sufficient— indication that our measures perform as intended. Future work should look into proper evaluation of the introduced metrics. The next opportunity to do so will be the INEX 2011 Data Centric Track on Faceted Search where our metrics will be used for evaluation. We expect that either or both of our evaluation metrics will foster the development of systems that focus on strategies of selecting the right facetvalues.

Acknowledgments. Maarten Marx acknowledges the financial support of the Future and Emerging Technologies (FET) programme within the Seventh Framework Programme for Research of the European Commission, under the FET-Open grant agreement FOX, number FP7-ICT-233599. This research was also supported by the Netherlands organization for Scientific Research (NWO) under project number 380-52-005 (PoliticalMashup).

References

Burke, R.D., Hammond, K.J., Young, B.C.: Knowledge-based navigation of complex information spaces. In: Proceedings of The National Conference On Artificial Intelligence, vol. 462, p. 468 (1996)

Dash, D., Rao, J., Megiddo, N., Ailamaki, A., Lohman, G.: Dynamic faceted search for discovery-driven analysis. In: Proceeding of the 17th ACM Conference on Information and Knowledge Mining, CIKM 2008, Napa Valley, California, USA, p. 3 (2008), doi:10.1145/1458082.1458087

English, J., Hearst, M., Sinha, R., Swearingen, K., Yee, K.P.: Hierarchical faceted metadata in site search interfaces. In: CHI 2002 Extended Abstracts on Human Factors in Computing Systems, pp. 628–639 (2002)

Hearst, M.: Design recommendations for hierarchical faceted search interfaces. In: ACM SIGIR Workshop on Faceted Search, p. 15 (2006)

Hearst, M.: Uis for faceted navigation: Recent advances and remaining open problems. In: Proc. 2008 Workshop on Human-Computer Interaction and Information Retrieval (2008)

Hearst, M.: Search user interfaces. Cambridge Univ. Pr., Cambridge (2009)

Järvelin, K., Kekäläinen, J.: Cumulated gain-based evaluation of IR techniques. ACM Transactions on Information Systems (TOIS) 20, 422–446 (2002) ACM ID: 582418

Kules, B., Capra, R., Banta, M., Sierra, T.: What do exploratory searchers look at in a faceted search interface? In: Proceedings of the 9th ACM/IEEE-CS Joint Conference on Digital Libraries, JCDL 2009, pp. 313–322. ACM, New York (2009); ACM ID: 1555452

Trotman, A., Wang, Q.: Overview of the inex, data centric track (2010)

Improving Query Expansion for Image Retrieval via Saliency and Picturability

Chee Wee Leong[1], Samer Hassan[1], Miguel Enrique Ruiz[2], and Rada Mihalcea[1]

[1] Department of Computer Science, University of North Texas
`cheeweeleong@my.unt.edu, samer@unt.edu`
[2] School of Library and Information Management, Emporia State University
`mruiz2@emporia.edu`

Abstract. In this paper, we present a Wikipedia-based approach to query expansion for the task of image retrieval, by combining salient encyclopaedic concepts with the picturability of words. Our model generates the expanded query terms in a definite two-stage process instead of multiple iterative passes, requires no manual feedback, and is completely unsupervised. Preliminary results show that our proposed model is effective in a comparative study on the ImageCLEF 2010 Wikipedia dataset.

1 Introduction and Motivation

The growth of the Internet has encompassed an enormous increase in the amount of data available in different modalities (e.g., texts, images, symbols, sound and video clips), formats (semi-structured vs unstructured), topics (e.g., politics, sports, entertainment) and languages (English is by far the most common). While different views of web data continued to emerge, the manner in which users specify queries remains largely unchanged. Typically, a user would enter one or more words in a natural language of choice, indicate the search domain (e.g., image search), and proceed to submit the query. Regardless of the type of web data requested, the length of the query is usually short, placing the onus on Information Retrieval (IR) systems to extrapolate beyond the surface forms of the query to extract its underlying semantics. Consequently, a good IR system must be intelligent enough to infer the true intentions of the user using further semantic analysis.

In this paper, we introduce a corpus-based approach of utilizing salient encyclopaedic concepts to expand queries for retrieving images. Our method is unsupervised, requires no manual feedback and involves a definite two-stage process to generate additional query terms that are semantically similar to the original query. We hypothesize that images are better represented using *picturable* words, and model this dimension of picturability in the expansion process using Flickr as a knowledge-base.

The paper is organized as follows. We briefly introduce how two resources, Wikipedia and Flickr, can be used to model the meaning and picturability of

P. Forner et al. (Eds.): CLEF 2011, LNCS 6941, pp. 137–142, 2011.

a word using *salient* concepts and corpus evidence respectively. After that, we proceed to construct our expansion model, and provide empirical results on a dataset showing its effectiveness.

2 Salient Semantic Analysis

We model the meaning of a word using its associated salient concepts that are linked in articles containing the word, using an approach termed Salient Semantic Analysis (SSA) [2]. The links within Wikipedia articles are regarded as clues or salient features (concepts) within the text that help define and disambiguate its context. By measuring the semantic association between words and salient concepts found in its neighborhood using co-occurrence statistics, SSA creates semantic profiles for words featuring the top concepts associated (co-occurring) with these words in a given window. Let us consider the following snippet extracted from a Wikipedia article:

> An *automobile*, *motor car* or *car* is a *wheeled* *motor vehicle* used for *transporting* *passengers*, which also carries its own *engine* or motor.

All the underlined words and phrases represent linked concepts, which are disambiguated and connected to the correct Wikipedia article. SSA semantically interpret each term in this example as a vector of its neighboring concepts (instead of words, as in other corpus-based measures). For example the word *motor* can be represented as a weighted vector of the salient concepts *automobile*, *motor car*, *car*, *wheel*, *motor vehicle*, *transport*, and *passenger*.

Formally, given a corpus C with m tokens, vocabulary size N, and concept size W (number of unique Wikipedia concepts), a $N \times W$ matrix (P) is generated representing the pairwise mutual information between each of the corpus terms with respect to its context concepts. The elements of P are defined as follows:

$$P_{ij} = log_2 \frac{f^k(w_i, c_j) \times m}{f^C(w_i) \times f^C(c_j)} \tag{1}$$

where f^k is the number of times the terms w_i and concept c_j co-occur together within a window of k words in the entire corpus.

To calculate the semantic relatedness between two words/texts, A and B, given the constructed matrix, we have:

$$Sim(A, B) = \begin{cases} 1 & Score_{cos}(A, B) > \lambda \\ Score_{cos}(A, B)/\lambda & Score_{cos}(A, B) \leq \lambda \end{cases} \tag{2}$$

where

$$Score_{cos}(A, B) = \frac{\sum_{y=1}^{N}(P_{iy} * P_{jy})^\gamma}{\sqrt{\sum_{y=1}^{N} P_{iy}^{2\gamma} * \sum_{y=1}^{N} P_{jy}^{2\gamma}}}, \tag{3}$$

The γ parameter allows the control of weight bias and λ is a normalization factor which help closing the semantic gap between perfect synonyms (tiger-tiger) and near-synonyms (tiger-feline).[1]

3   Flickr Picturability

We hypothesize that some words are more *picturable* than others (e.g. banana vs paradigm), i.e., it is easier to find an image to visually describe concepts invoked by the more picturable word. In our work, we attempt to model the picturability of a word using Flickr[2] as a resource.

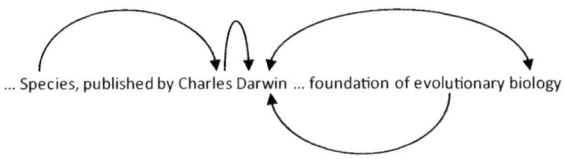

Fig. 1. Flickr Picturability Labels

Following [3], we can model the picturability of a word alone, or in association with other words in text. The latter is of particular interest since we wish to discriminate words based on their picturability in order to observe the corresponding effects on the performance of query expansion models. The algorithm proceeds as follows: given a word in a free text, we use the *getRelatedTags*[3] API to retrieve the most frequent Flickr tags associated with the word, and use them as corpus evidence to compute its picturability score. We disregard stopwords and any word less than three characters long or not found in Flickr tag repository. Next, any retrieved tag appearing as surface forms in the text is rewarded proportionally to its term frequency in the text, as words having a high frequency and featuring more in the Flickr tags are relatively more picturable in the text. Additionally, we score all words that return a non-empty related tags set with a discounted weight ($\beta=0.5$) of its term frequency to promote outstanding, picturable words in the rewarding phase. As an illustration, consider the following snippet:

> *On the Origin of Species, published by Charles Darwin in 1859, is considered to be the foundation of evolutionary biology.*

For each of the content words w_i (i.e. *origin, species, published, charles, darwin, foundation, evolutionary,* and *biology*), we query Flickr and obtain their related tag set R_i. The words *origin, published,* and *foundation* return an empty set of related tags and hence are not scored and also removed from our set of consideration, leaving *species, charles, darwin, evolutionary,* and *biology* with the

---

[1] In a sense, λ represents a synonymy threshold, a value above which the semantic relationship is considered synonymous.

[2] http://www.flickr.com

[3] http://www.flickr.com/services/api/flickr.tags.getRelated.html

initial score of 0.5. Next, we score each w_i based on the number of votes it receives from the remaining w_j (Figure 1). Each vote represents an occurrence of the candidate tag w_i in the related tag set R_j of the candidate tag w_j. For example, if *darwin* appeared in the Flickr related tags for *charles*, *evolutionary* and *biology*, then it would have a weight of 0.5+1+1+1=3.5.

4   Automatic Query Expansion via Saliency and Picturability

In this work, we focus on the task of query expansion for retrieving images. Our model entails a two-stage process, namely, candidates generation and candidates selection, described below.

4.1   Candidate Generation

Given a set of initial query terms $Q=\{q_1,q_2,...,q_k\}$, we retrieve a list of the top m Wikipedia articles most relevant to Q, computed by summing over individual concept vectors of all query terms in Q using SSA, as described in section 2. A pre-processing step is first applied to remove stopwords from the title of each article. We consider each of the remaining words w_i a candidate for query expansion, weighted using a simple fusion rule:

$$Weight(w_i) = tf(w_i) * 1/rank(w_i) * flickr(w_i) \qquad (4)$$

where $tf(w_i)$ is the term frequency of w_i appearing as a word across all m Wikipedia titles, $rank(w_i)$ is the highest rank of the title (reverse sorted) containing w_i, and $flickr(w_i)$ is picturability score provided by corpus evidence using the Flickr picturability method in section 3.

4.2   Candidate Selection

From the generation phase, we select as working set the top ranked W words according to (4) which can be potentially applied to expand Q. We next adopt a bootstrapping procedure to expand Q as follows: for each word w_j traversed in W (reverse sorted), if $Sim(Q,w) \geq \alpha$, we update Q to include w_j, where $Sim(Q,w)$ is provided by SSA in section 2. Our expansion model focuses on extracting picturable terms associated with salient concepts that are semantically similar to the initial query text Q. The weighting scheme in the candidate generation phase ensures that terms strongly associated with these concepts are first used to expand Q, and the expansion is performed incrementally, adding the most relevant terms each time while preserving the overall semantic consistency.

5   Empirical Evaluation

For evaluating our expansion model, we use the data from the ImageCLEF 2010 Wikipedia [6]. This collection includes 237,434 images with associated texts in

English, French and German. Approximately 10% of these images are annotated in all three languages, 24% with annotations in two languages, and 62% with annotations in one language. For each image used in the evaluation, we translated the French and German texts in the captions into English using the Bing Translation service. The collection also contains 70 topics written in all three languages. For a fair comparative study, we build our retrieval system using the same specifications provided by the best performing system using exclusively monolingual features (untaTxEn) [7]. Following their approach, we first index all data in the collection using the Indri/Lemur information retrieval system. Next, we build a unigram model with Dirichlet smoothing, Krovetz stemming and a list of English stopwords [5]. During retrieval, each retrieved document, D, is scored as follows:

$$P(Q|D) = \prod_i P(q_i|D)^{\frac{1}{|Q|}} \tag{5}$$

where Q is the query in question and q_i is a query term in Q. Prior to retrieval, we perform query expansion for each query Q using the expansion model explained in section 4, with $m=1000$ to ensure adequate coverage of concepts, $W=50$ and $\alpha = 1$. When measuring similarity, our SSA model is set to $\gamma = 1.2$ and $\lambda = 0.02$ which are derived from experiments using three manually constructed queries. The topics can be classified into three different tiers of difficulty based on results from teams participating in ImageCLEF. Here, we show examples of automatically expanded query for each tier. {*cockpit of an airplane*} yields {*cockpit airplane compartment aircraft flight plane airplane passenger carrier boeing crash*} (easy), {*dna helix*} yields {*dna helix strand fold protein molecular alpha binding*} (medium), while {*building site*} yields {*building site construction structures*} (hard).

6   Discussion

Table 1 shows the results from our experiments. The performance of each system is reported using a collection of metrics typically used in IR. As observed, our query monolingual expansion model SSA (flickr + ENG) generally yield improvements over the top-performing monolingual querying system (untaTxEn) in ImageCLEF 2010, scoring better in number of relevant images retrieved, MAP and bpref, and is competitive (0.2916 vs 0.3025) on the Rprec metric. The difference in retrieval performance between SSA (flickr+EN) and SSA (flickr+EN+DE+FR) represents the advantage of querying in a multilingual setting, which records improvements in all metrics except for number of relevant images retrieved. Similarly, the difference between SSA (EN+DE+FR) and SSA (flickr+EN+DE+FR) indicates the role played by the flickr picturability component in equation 4. When omitted, the flickr picturability causes a drop in retrieval performance recorded in all metrics. Overall, with the exception of Precision at 20, our expansion models (monolingual or multilingual) scores significantly better performance over the average system participating in CLEF 2010.

Table 1. Retrieval Performance

System	#Relevant	Map	Rprec	bpref	P@20
SSA (flickr + EN)	**8176**	0.2277	0.2916	0.2654	0.3314
SSA (EN+DE+FR)	8057	0.2240	0.2900	0.2655	0.3871
SSA (flickr + EN+DE+FR)	8162	**0.2293**	0.2971	**0.2685**	**0.4114**
untaTxEn	7840	0.2251	**0.3025**	0.2617	0.4057
Average in CLEF2010	5789	0.1611	0.2323	0.2032	0.3582

7   Related Work

Several expansion models based on Wikipedia have sprung into existence recently [1,4,8]. The work most closely related to ours is [1], where each query is run using a dependence model to retrieve a list of ranked Wikipedia documents, of which anchor words in lower-ranked documents referencing higher-ranked ones are utilized to expand the query. In contrast, our system is different in a number of aspects. First, we are adopting a lightweight approach that performs word mining from the surface forms of articles titles, rather than analysing entire documents for anchor words. Second, we employ an additional level of semantic relatedness check using SSA to ensure the words in the expanded query are semantically consistent. Finally, our system preferentially selects words that are not only semantically similar to the original query, but also picturable ones. As future work, we plan to address the optimization of system parameters.

References

1. Arguello, J., Elsas, J.L., Callan, J., Carbonell, J.G.: Document representation and query expansion models for blog recommendation. In: Proceedings of the Second International Conference on Weblogs and Social Media (2008)
2. Hassan, S., Mihalcea, R.: Semantic relatedness using salient semantic analysis. In: Proceedings of AAAI Conference on Artificial Intelligence (2011)
3. Leong, C.W., Mihalcea, R., Hassan, S.: Text mining for automatic image tagging. In: Proceedings of the International Conference on Computational Linguistics (2010)
4. Li, Y., Luk, W.P.R., Ho, K.S.E., Chung, F.L.K.: Improving weak ad-hoc queries using wikipedia as external corpus. In: Proceedings of ACM SIGIR Conference on Research and Development in Information Retrieval (2007)
5. Ogilvie, P., Callan, J.: Experiments using the lemur toolkit. In: Proceedings of 2001 Text REtrieval Conference (TREC 2001). National Institute of Standards and Technology (2001)
6. Popescu, A., Tsikrika, T., Kludas, J.: Overview of the wikipedia retrieval task at imageclef 2010. In: CLEF 2010 Labs and Workshops, Notebook Papers. Padua, Italy (2010)
7. Ruiz, M., Chen, J., Pasupathy, K., Chin, P., Knudson, R.: Unt at imageclef 2010: Clir for wikipedia images. In: CLEF 2010 Labs and Workshops, Notebook Papers. Padua, Italy (2010)
8. Xu, Y., Jones, G.J., Wang, B.: Query dependent pseudo-relevance feedback based on wikipedia. In: Proceedings of the ACM SIGIR Conference on Research and Development in Information Retrieval (2009)

Author Index